Invasive Plants of West Virginia, USA; the Philippines; and Malaysia:

A Collaborative Online International Learning (COIL) Project

Contributing Editor, Dr. Rico M. Gazal

Contributors, Dr. Mary Ann R. Agudilla, Dr. Amalia E. Almazol, Dr. Alona C. Linatoc, Dr. Aqilah Awg Abdul Rahman

Sabre Lake Publishing

2026

Invasive Plants of West Virginia, USA; the Philippines; and Malaysia: A Collaborative Online International Learning (COIL) Project

No artificial intelligence (AI) was used in the production of this book.

Published in the United States of America

Production Design by Melissa Gish
Sabre Lake Publishing
Minnesota

ISBN 978-1-953033-21-5

10 9 8 7 6 5 4 3 2 1

Table of Contents

Note to Readers

Invasive plants are not simply local problems. They are interconnected ecological challenges shaped by global movement, regional land use, and ecosystem dynamics. When students and faculty commit to collaboration, meaningful learning can cross borders. Collaborative Online International Learning (COIL) makes that commitment possible by connecting students and faculty across countries through shared online course activities integrated directly into coursework. This book documents one such project, carried out across four institutions in three countries: Glenville State University in West Virginia, United States; the University of the Philippines Los Baños and Southern Luzon State University in the Philippines; and Universiti Tun Hussein Onn Malaysia in Malaysia.

The book is organized in several parts. Opening chapters written by faculty provide conceptual foundation in invasion ecology, the design and rationale of the COIL module, and related topics including forest systems, shared traits of invasive species, and management approaches. These are followed by student essays that combine field observation with research. The essays are organized by region and illustrate how invasion ecology plays out differently across West Virginia, the Philippines, and Malaysia. They reflect careful observation, curiosity, and a growing sense of environmental awareness. The essays model the kind of research and critical thinking the project is designed to develop. Moreover, they demonstrate that higher education can connect learners across cultures, strengthen communication and teamwork, and explore environmental problems through both local knowledge and global perspectives that matter for conservation, forestry, and natural resource management.

Closing chapters by partner faculty reflect on these topics as well as the lessons learned from implementing this collaboration. Readers who want to understand the full structure of this COIL project, including all activities, discussion topics, and the timeline used to coordinate the four institutions, will find it in Appendix A. Appendix A also includes the semester calendar and course descriptions used in this project. These are provided in enough detail that faculty at other institutions can adapt them to their own course structures and institutional calendars. Readers planning their own COIL projects will find it useful to review the calendar in Appendix A alongside the "Learning Together Through COIL" chapter, which explains the pedagogical reasoning behind each phase of the project. Faculty considering a COIL project will find this book most useful as a working model rather than a rigid template. The course descriptions show how the learning objectives for invasion ecology and intercultural competence were framed across different disciplines and can serve as starting points for adapting the project to other ecology or natural resource courses.

This book demonstrates that international collaboration in science education is not only achievable but successful across significant differences in geography and resources. The project described here reflects the collaborative efforts of faculty who co-designed the learning experience and students who participated with strong commitment and engagement. May this book serve as a resource for future students and educators, and as a reminder that collaboration is not only a teaching strategy, but also a scientific practice. We hope readers will take what is useful here and carry it into their own teaching and learning.

Rico M. Gazal, PhD

Preface

This book began with a simple teaching goal: helping students learn invasion ecology by studying invasive plants in their own regions and comparing their observations with students in other countries. Invasive plants are well suited for this type of learning because they are visible on the landscape and raise questions with both local and global importance. Why do some species spread rapidly in one environment but not in another? How do disturbance, climate, soils, and human activity influence invasion patterns?

To support this goal, we developed an 11-week Collaborative Online International Learning (COIL) experience that connected forest ecology and biodiversity courses across four institutions in three countries: Glenville State University in West Virginia, United States; the University of the Philippines Los Baños College of Forestry and Natural Resources and Southern Luzon State University in the Philippines; and Universiti Tun Hussein Onn Malaysia in Malaysia. Faculty partners worked together to plan the learning activities and coordinate student collaboration despite distance and time differences.

The learning goals were both ecological and collaborative. Students examined invasive plant ecology, introduction pathways, and environmental impacts in their regions. They shared observations through discussion topics and short videos and created concept maps to explain how invasive plants interact with soil, climate, vegetation, and landscape conditions. The module concluded with a written report and presentation, followed by peer viewing and reflection.

In addition to the student essays, faculty members contributed chapters that provide broader context for understanding invasion ecology and the learning experience. These chapters discuss topics such as the ecology of biological invasions, learning together through COIL, viewing forests as interconnected systems, understanding shared traits of invasive species, management approaches, and lessons learned from implementing this international collaboration.

The chapters in this book are written as reflective, research-informed essays rather than formal scientific papers. Each essay combines field observations and credible sources to explain invasive plants and ecosystem dynamics. The goal is to share learning across countries, highlight student perspectives, and encourage comparison and discussion.

The complete module designed to implement COIL on *Understanding the Impacts of Invasive Species and Ecosystem Dynamics* is provided in Appendix A, including the project overview and timeline used to guide the collaborative learning activities.

These essays reflect careful observation, research, and systems thinking. They also show how comparing invasive species across countries can deepen understanding of ecosystem processes. I am grateful to our faculty partners for their collaboration and to our students for their professionalism in completing this work.

Acknowledgments

This book is the result of a shared commitment to learning and collaboration across three countries. We extend our appreciation to the faculty, students, and administrators of our four partner institutions: Glenville State University (West Virginia, USA); the University of the Philippines Los Baños College of Forestry and Natural Resources; Southern Luzon State University (Philippines); and Universiti Tun Hussein Onn Malaysia. Your time, support, and willingness to work together made it possible to implement this COIL module and bring this student essay collection to publication.

We also gratefully acknowledge the GSU Foundation and the EQT Foundation for financial support for the reproduction and launch of this book. This support helped transform an instructional project into a lasting resource for students, educators, and partners.

Special thanks are due to Glenville State University's Office of Technology staff for their assistance in setting up and supporting Brightspace as the module's online course platform. Your technical guidance and responsiveness were essential in delivering a smooth and accessible international learning experience for all participants.

We also thank Dr. Megan Gibbons, who pioneered COIL at Glenville State University and helped spearhead the international partnerships that continue to strengthen our global learning efforts.

Contributors

Mary Ann R. Agudilla, PhD is a Professor III and Chairperson of the Forestry Department at Southern Luzon State University (SLSU) in Lucban, Quezon, Philippines, where she has been a faculty member since 1995. She holds a PhD in Forestry from the University of the Philippines Los Baños, earned with support from the DOST Accelerated Science and Technology Human Resource Development Program. A licensed forester who placed 10th in the national board examination, her research centers on biodiversity assessment, insect ecology, watershed conservation, and ecosystem services valuation. Her work has been published in journals including the *Philippine Journal of Science* and the *Asian Journal of Biodiversity*, and she has presented at numerous national and international conferences. She is a member of the Forest and Natural Resources Research Society of the Philippines and the Philippine Native Plants Conservation Society.

Amalia E. Almazol, PhD is a Professor VI in the College of Forestry and Natural Resources at Southern Luzon State University (SLSU) in Lucban, Quezon, Philippines, where she has served on the faculty since 2001. A licensed forester, she holds a PhD in Botany (major in Ecology) from the University of the Philippines Los Baños and a Master of Science in Forestry from De La Salle-Araneta University, where her thesis was recognized as Best in its category. Her research spans forest botany, mangrove ecology, pollination biology, biodiversity assessment, and apiculture, with sustained focus on the ecosystems of Mt. Banahaw and the Pagbilao Mangrove Forest. She has served as Chairperson of the BS and Master in Forestry Programs and as Dean of the College of Agriculture, and received the 2023 Outstanding Regional Forester of the Year Award for Region IV-A. She was awarded Outstanding Professional in Forestry by the Philippine Professional Regulation Commission in 2024.

Rico M. Gazal, PhD is a Professor of Forestry in the Department of Land Resources at Glenville State University in Glenville, West Virginia, where he has taught since 2005 and served multiple terms as Department Chair. A West Virginia Licensed Professional Forester, he holds a PhD in Forest Resources (Tree Physiology) from Mississippi State University and both his bachelor's and master's degrees in Forestry from the University of the Philippines Los Baños, where he graduated cum laude. His teaching spans forest ecology, forest measurements and management, GIS applications, and drone technology, and he has supervised more than 50 undergraduate research projects. His scholarly contributions include peer-reviewed publications and forthcoming chapters on tree biology and physiology for an IUFRO open textbook. He is a recipient of the 2025 Faculty Award of Excellence at Glenville State University and was a finalist for West Virginia Professor of the Year.

Alona C. Linatoc, PhD is an Associate Professor in the Department of Forest Biological Sciences at the College of Forestry and Natural Resources, University of the Philippines Los Baños (UPLB), where she also serves as Director of the Office of Counseling and Guidance and Director of the Training Center on Tropical Resources and Ecosystems Sustainability. A Registered Professional Forester and Certified Global Biodiversity Standard Assessor, her expertise spans forest ecology, forest physiology, forest botany, and biodiversity conservation. She holds a PhD in Forest Physiology from Universiti Malaysia Sabah and has published widely in indexed journals, with a Google Scholar h-index of 11 and over 328 citations. She serves on the editorial boards of several international journals and is a member of the National Research Council of the Philippines, IUFRO, and the Society of Conservation Biology.

Aqilah Awg Abdul Rahman, PhD is a Senior Lecturer and Researcher in the Faculty of Applied Sciences and Technology at Universiti Tun Hussein Onn Malaysia (UTHM), where she also serves as Head of the Centre of Research for Sustainable Uses of Natural Resources. She holds a PhD in Entomology/Biodiversity and Conservation from UTHM and a Bachelor of Science in Molecular Bioscience and Biotechnology from the Rochester Institute of Technology, where she graduated with highest distinction. Her research focuses on biodiversity and conservation, entomology, species distribution modeling, and ecosystem carbon regulation. She has secured multiple competitive research grants as principal investigator, with projects centered on butterfly ecology and ecosystem health assessment in tropical forest landscapes. She is a Council Member of the Entomological Society of Malaysia and a member of the Society for Conservation Biology, Malaysia Chapter.

Melody A. Wise, PhD is an Associate Professor of English and English Education Specialist in the Department of Language and Literature at Glenville State University, where she also serves as Assessment Coordinator and Writing Center Director. She holds a PhD in English with a specialization in Composition and TESOL from Indiana University of Pennsylvania. Her teaching encompasses writing, literature, and English Education methodology, with expertise in multimodal and technology-enhanced approaches to literacy instruction. Her scholarly work addresses writing program administration, assessment, and multiliteracies, and she has presented at national conferences including the Conference on College Composition and Communication. She is a member of the National Council for Teachers of English and the Council of Writing Program Administrators.

PART I

Shared Foundations

Rico M. Gazal, PhD

01

Ecology of Invasion

Ecology of Invasion

Today, our forest ecosystems face numerous challenges, ranging from localized issues to global concerns, including the proliferation of invasive species. Invasive species are organisms such as plants, animals, insects, and microbes that are introduced accidentally or deliberately to an area where they are not native. While many introduced species fail to establish in new environments, a small fraction becomes invasive, forming self-sustaining populations, spreading across landscapes, and causing ecological or socioeconomic harm (IPBES, 2023).

Invasion is best described in stages. First comes arrival, when seeds or plant fragments enter an area through trade, travel, or deliberate planting. Next is establishment, when those propagules survive local conditions and reproduce. Then comes spread, as populations expand and disperse to new sites. More introduction events and larger numbers of individuals increase the chance that at least some will reach suitable habitats, survive the early establishment phase, and persist long enough to reproduce. These stages can unfold over a few years or extend across decades or even centuries. Alien species are being recorded worldwide at increasing rates as global trade and travel move organisms between regions more often and from a wider range of places (Seebens et al., 2018).

There are several ways a new species can be introduced into an area, and these human actions shape what happens in ecosystems. Horticulture is a major source of new plants because it intentionally moves species with traits people value, such as fast growth, hardiness, tolerance of stress, and in some cases the ability to thrive in disturbed or nutrient poor soils, traits that can also aid establishment in the wild (IPBES, 2023). Other introductions occur unintentionally through contaminants in nursery stock and soil, seeds mixed with grain and animal feed, plants carried on vehicles and machinery, and long-distance transport through shipping and trade. These routes of entry can reach even the places we most want to protect, making their management increasingly critical as climate and land use change reshape which introductions succeed (Lieurance et al., 2025).

Once a plant arrives, its survival depends on the conditions in the new environment. Disturbance often creates the opportunity an invader needs to establish. Fire, storms, logging, road building, and cultivation can reduce native cover and increase light, moisture, and nutrients, creating pulses of resources that favor rapid colonizers. Many invasions are therefore about timing and opportunity. A plant that arrives just after a disturbance can gain a head start, and that early advantage can build over time as it grows larger, captures more light, and produces more seeds.

A new plant's ability to compete can lead to dominance over time. Compared with many native species, invasive plants often show trait combinations linked to rapid resource capture and fast growth. Many invaders have higher specific leaf area and related traits associated with rapid carbon gain (Ordonez et al., 2010). However, invasion success is rarely explained by a single trait. Instead, it reflects how multiple traits work together across life stages, including high reproduction, adaptable growth responses to local conditions, efficient dispersal, and strong performance across a wide range of environments (Kaushik et al., 2022).

Impacts go beyond the plants we see. As invasive plants spread, they can alter site conditions and threaten native plant species, forest productivity, and long-term ecosystem stability. Their success often originates from strong competitive abilities, including rapid growth, prolific seed production, and in some cases allelopathy, release of chemicals that suppress the growth of co-existing plants. Some invaders also modify soil microbial communities and nutrient cycling, creating conditions that further favor their own growth. Chemical changes to soils and the release of inhibitory compounds from stems, leaves, and roots can help invasive plants maintain dominance, especially where disturbance increases resource availability such as sunlight, soil moisture, and nutrients. Invasive alien plants often reduce the abundance and diversity of native species and can alter community structure and ecosystem processes (Vilà et al.,2011). Recent studies warn that invasive alien species are a major direct driver of biodiversity loss and can affect ecosystem services and human well-being (Pyšek et al., 2020; IPBES, 2023). Because biodiversity supports ecosystem functioning and stability, invasions that simplify communities can alter fire regimes and hydrology, reduce wildlife habitat quality, and shift carbon and nutrient storage (Tilman et al., 2014).

For students writing about invasive plants in the United States, the Philippines, and Malaysia, the key lesson is that invasions are not accidents of biology alone. They are shaped by human movement, ecological opportunity, and interactions within particular ecosystems. A plant's arrival explains how it was introduced, local ecology explains why it establishes and spreads, and its impacts explain why land managers are concerned. Prevention and early detection are most effective before spread and damage become difficult to reverse, and projections suggest alien species will continue accumulating through mid-century without stronger controls on introductions (Seebens et al., 2020).

References

Cardinale, B. J., Duffy, J. E., Gonzalez, A., Hooper, D. U., Perrings, C., Venail, P., Narwani, A., Mace, G. M., Tilman, D., Wardle, D. A., Kinzig, A. P., Daily, G. C., Loreau, M., Grace, J. B., Larigauderie, A., Srivastava, D. S., & Naeem, S. (2012). *Biodiversity loss and its impact on humanity. Nature, 486*(7401), 59–67.

Intergovernmental Science Policy Platform on Biodiversity and Ecosystem Services. (2023*). Thematic assessment report on invasive alien species and their control.* In H. E. Roy, A. Pauchard, P. Stoett, & T. Renard Truong (Eds.). IPBES secretariat.

Kaushik, P., Pati, P. K., Khan, M. L., & Khare, P. K. (2022). *Plant functional traits best explain invasive species' performance within a dynamic ecosystem: A review. Trees, Forests and People, 8,* 100260.

Lieurance, D., Canavan, S., Faulkner, K. T., O'Shaughnessy, K. A., Lockwood, J. L., Parsons, E. W., Avery, J. D., & Daniel, W. M. (2025). *Understanding and managing introduction pathways into protected areas in a changing climate. Biological Invasions, 27*, 74.

Ordonez, A., Wright, I. J., & Olff, H. (2010). *Functional differences between native and alien species: A global-scale comparison. Functional Ecology, 24*(6), 1353–1361.

Pyšek, P., Hulme, P. E., Simberloff, D., Bacher, S., Blackburn, T. M., Carlton, J. T., Dawson, W., Essl, F., Foxcroft, L. C., Genovesi, P., Jeschke, J. M., Kühn, I., Liebhold, A. M., Mandrak, N. E., Meyerson, L. A., Pauchard, A., Pergl, J., Roy, H. E., Seebens, H., & Richardson, D. M. (2020). *Scientists' warning on invasive alien species. Biological Reviews, 95*(6), 1511–1534.

Seebens, H., Blackburn, T. M., Dyer, E. E., Genovesi, P., Hulme, P. E., Jeschke, J. M., Pagad, S., Pyšek, P., van Kleunen, M., Winter, M., & Essl, F. (2018). *Global rise in emerging alien species results from increased accessibility of new source pools. Proceedings of the National Academy of Sciences, 115*(10), E2264–E2273.

Seebens, H., Bacher, S., Blackburn, T. M., Capinha, C., Dawson, W., Dullinger, S., Genovesi, P., Hulme, P. E., van Kleunen, M., Kühn, I., Jeschke, J. M., Lenzner, B., Liebhold, A. M., Pergl, J., Pyšek, P., Winter, M., & Essl, F. (2020). *Projecting the continental accumulation of alien species through to 2050. Global Change Biology.* Advance online publication.

Tilman, D., Isbell, F., & Cowles, J. M. (2014). *Biodiversity and ecosystem functioning. Annual Review of Ecology, Evolution, and Systematics, 45,* 471–493.

Vilà, M., Espinar, J. L., Hejda, M., Hulme, P. E., Jarošík, V., Maron, J. L., Pergl, J., Schaffner, U., Sun, Y., & Pyšek, P. (2011). *Ecological impacts of invasive alien plants: A meta-analysis of their effects on species, communities and ecosystems. Ecology Letters, 14*(7), 702–708.

Rico M. Gazal, PhD

02

Learning Together Through COIL

Learning Together Through COIL

We developed this book to compile student essays on invasive plants from their own countries, using Collaborative Online International Learning (COIL) to help students learn from one another through shared online activities. COIL brings students and faculty across cultures into a common set of course activities where they learn, discuss, and collaborate in ways that make global learning accessible beyond traditional travel (SUNY, 2025). In practice, COIL works best when the activities and deadlines are fully integrated into the course and students have meaningful interactions with international peers.

In this project, the COIL module is an 11-week collaboration that links Forest Ecology and related biodiversity courses across four institutions in three countries: Glenville State University in West Virginia, United States; the University of the Philippines Los Baños College of Forestry and Natural Resources and Southern Luzon State University in the Philippines; and Universiti Tun Hussein Onn Malaysia in Malaysia. Rather than treating invasive plants as a topic that belongs to one region, the module invites students to ask a bigger ecological question: how do species traits, human pathways, and local ecosystem conditions interact to shape invasion success and ecosystem responses in different places.

The learning experience follows a progression that COIL practitioners often recommend, moving from community building to comparative inquiry, then to a shared product, and finally to presentation and reflection (SUNY, 2025). Students begin by introducing themselves and establishing expectations for respectful collaboration and academic integrity. From there, the course shifts into its main focus. Groups select an invasive species and develop a shared understanding of its identity, habitat, and impacts through discussion boards and field-based "spotlight" videos that emphasize identification and the environmental conditions that support growth and spread.

A key feature of the module is that students go beyond identification and description to analyze the interactions between an invasive plant and its environment. They build an explanatory model of how the invasion works, and the concept map assignment is key to this analysis. Students connect the invasive species to ecosystem components such as soil; climate, including temperature, humidity, and precipitation; biota, including plants, animals, insects, and other living components; and physiography, including topography, slope, elevation, and related factors. This turns individual facts into systems thinking and highlights the connections that drive ecological outcomes, such as how disturbance increases light and soil resources, how soil microbes shift in ways that favor an invader, how dispersal is often driven by human activity, how interactions with animals and insects can influence spread, and how management actions can change competitive outcomes. When students map these interactions, they can see why a species becomes invasive in one setting but not in another.

International comparison deepens this learning in a way that a single site case study cannot. Students in West Virginia may recognize invasion pressures tied to roads, timber harvest, and recreation. Students in the Philippines may bring examples shaped by agriculture, typhoons, or rapid land use change. Students in Malaysia may frame invasions alongside plantation landscapes, forest fragments, and biodiversity conservation priorities. When these perspectives come together, the same invasive strategy can look different because the ecological context is different. The central lesson in ecosystem dynamics is that outcomes emerge from interactions, not from one factor alone.

The collaborative structure of this COIL project also makes the experience authentic. The module culminates in a group-written report and an oral presentation, requiring students to synthesize the factors driving invasion for their species within their region. Because time zones are a real constraint, the module intentionally mixes synchronous and asynchronous participation, using recorded presentations and shared viewing to ensure everyone can contribute meaningfully. Students then watch peer presentations from partner institutions and post reflections that include comparisons, questions, and feedback, reinforcing learning beyond the final submission.

This design also matches what recent research suggests about effective COIL. Faculty from all three countries co-planned the learning experience, shared outcomes, and coordinated the sequence of activities, even though most final group products were completed by students within the same country because of time zone constraints. Intercultural learning in this module happens through structured exchanges, including discussion topics that compare invasive plants across countries and regions, peer feedback on spotlight videos, comments on the complex interactions captured in concept maps, and reflections after watching recorded presentations from partner institutions. Pre and post surveys were used to document baseline understanding and capture perceived growth in ecological learning and intercultural competence. COIL practice emphasizes the value of co-designed learning and structured activities that promote intercultural engagement through shared tasks, discussion, and reflection (Hackett et al., 2024). Empirical studies also report gains in intercultural competence in COIL and related virtual exchange formats, especially when students have structured interaction, guided tasks, and reflection opportunities (Chu et al., 2024; Hackett et al., 2023). In this project, invasive plants work especially well because this topic naturally demands comparison across places, showing how human-mediated introduction and local ecological conditions shape invasion outcomes.

This chapter is about learning together as a scientific practice. COIL turns invasive plants into a shared ecological problem that spans landscapes. By working across the United States, the Philippines, and Malaysia, students see invasion ecology as both local and global at the same time, and they practice collaborative inquiry that today's environmental challenges require).

References

Chu, C., Hooper, T., Takahashi, M., & Herke, M. (2024). The effect of asynchronous virtual exchange on intercultural competence among undergraduate students. *Journal of Comparative & International Higher Education*, *16*(2), 213–226.

Hackett, S., Dawson, M., Janssen, J., & van Tartwijk, J. (2024). Defining collaborative online international learning (COIL) and distinguishing it from virtual exchange. *TechTrends*, *68*, 1078–1094.

Hackett, S., Janssen, J., Beach, P., Perreault, M., Beelen, J., & van Tartwijk, J. (2023). The effectiveness of collaborative online international learning (COIL) on intercultural competence development in higher education. *International Journal of Educational Technology in Higher Education*, *20*, 5.

Rubin, J. (2017). Embedding collaborative online international learning (COIL) at higher education institutions: An evolutionary overview with exemplars. *Internationalisation of Higher Education*, *2*, 28–44.

State University of New York, SUNY. (2025). *What is COIL?* Retrieved December 20, 2025, from https://online.suny.edu/introtocoil/suny-coil-what-is/.

Alona C. Linatoc, PhD

03

Seeing the Forest as a System

Seeing the Forest as a System

Introduction

Forests are widespread, covering nearly 30% of the Earth's land surface (SCBD, 2001) and typically have a predominant species composition. Tropical forests are found near the equator, between the Tropic of Cancer (23.5° N) and the Tropic of Capricorn (23.5° S). This region receives overhead sunlight year-round; hence, it is characterized by a consistently warm climate (>20°C/68°F) that is suitable for a myriad of flora and fauna to thrive. Within the tropical region are different forest ecosystems such as tropical rainforests, tropical dry or monsoon forests, and mangrove forests. Above and below the tropical region, ~30-50° north and south, respectively, are temperate regions comprised of temperate deciduous forests, temperate coniferous forests, and temperate rainforests. Higher up in the temperate region, located between 50-60° north, are boreal forests, or taiga (note that there is no taiga towards the south pole). Between the tropical region and the taiga are a few Mediterranean forests, characterized by hot, dry summers and mild, wet winters.

Forest Strata

A forest is composed of several strata, each with a unique microclimate that provides habitat for a specific group of organisms. These strata illustrate how thousands of different species coexist in microhabitats within a limited space.

Typical of tropical rainforests are the emergent trees belonging to the families Dipterocarpaceae, Malvaceae, Fabaceae, and Lecythidaceae. Emergent trees reaching up to 100 m (328 ft), are individual trees scattered, rising above the main canopy that receive full sunlight and are characterized by winged, wind-dispersed fruits (samara). Some organisms associated with emergent trees include epiphytes (e.g., orchids), bats, eagles, hawks, and insects (e.g., bees and butterflies). Below the emergent trees is a continuous cover - the main canopy, which captures most of the sunlight, thus regulating the microclimate conditions underneath. This layer of the forest provides food and shelter for diverse plant species (e.g., epiphytes, lianas, climbers) and a wide range of animals, from mammals to birds to reptiles and insects. Below the main canopy is the understory, which receives much less light (~50%) than the main canopy. Shade-tolerant species of shrubs, saplings, and herbaceous plants dominate this layer. Species in the families Rubiaceae, Moraceae, Euphorbiaceae, and Palmae (Arecaceae) are some of the main food sources for small mammals, reptiles, and insects. The forest floor is the lowest layer and receives the least sunlight (~25%), where decomposition and nutrient cycling happen. Associated organisms in this layer include detrivores (e.g., earthworms, dung beetles, millipedes), fungi, and decomposing materials such as leaf litter, wood, and dead animals.

It is common for large trees to fall due to a rotten trunk or strong winds. This creates significant gaps that allow pioneer (sun-tolerant), fast-growing species to flourish, providing additional food for wildlife and sustaining the diversity of life in the forest. Over time, as the gap closes, pioneer species create environmental conditions that support the growth of shade-tolerant emergent trees.

Forest Dynamics

A forest is a dynamic ecosystem built on the complexities of interactions among biotic and abiotic components. The products and services are derived from the exchanges and processes of the multi-dimensional biophysical systems. This function involves all ecological and evolutionary processes, including gene flow, disturbances, and biogeochemical cycling. Abiotic factors, such as elevation, slope, soil characteristics, and climate (temperature, rainfall, sunlight), comprise the physical environment and the main influences on species distribution. These parameters become the primary factors for which biological diversity adapts. The association of different species forms relationships that shape an ecosystem. Every species in the ecosystem has either beneficial (mutualism, protocooperation, facultative, commensalism, or obligate) or antagonistic (allelopathy, herbivory, predation, or competition, ammensalism) relationships. These relationships maintain ecosystem balance and, to a certain extent, keep the ecosystem in equilibrium.

Ecological Services of Forests

Globally, forest types are defined by climate, much as forests regulate local and global climate, ameliorate weather events, regulate the hydrological cycle, protect watersheds and their vegetation, water flows, and soils, and provide a vast store of genetic information, much of which has yet to be uncovered.

Through photosynthesis, trees and other plants store billions of tons of carbon, thereby mitigating the greenhouse effect. Estimates reveal that at least 860 gigatons of carbon are absorbed in the world's forests, of which 44% in soil stored up to 1 m in depth, 42% in the live above- and below-ground biomass, 8% in dead wood, and 5% in litter. Tropical forests store most of their carbon in the vegetation (biomass), whereas boreal forests store significant amounts of carbon in soils (Pan et al., 2011).

Forests are vital in providing clean water to millions of people worldwide. During rainfall, as water is intercepted by canopies and the forest floor, layers of leaf litter and porous soil trap sediment and particles, effectively filtering water as it flows through streams and rivers. The roots of trees not only anchor the soil to prevent erosion but also act as sponges, slowing the impact of heavy rainfall and reducing flash floods. During dry periods, the stored groundwater is slowly released into the river, maintaining its flow year-round. Through transpiration, vegetation releases water vapor into the atmosphere, significantly contributing to cloud formation and thus playing a major role in the hydrologic cycle. Moreover, the presence of different forest strata filters sunlight, providing shade vital for regulating temperature and supporting life.

Forests are more than the sum of their biotic and abiotic components. The forest's physical structure serves as a crucial natural shield against storms/typhoons, breaking strong winds and waves, and protecting countless lives and property from damage. Intact forests at higher elevations protect against landslides and erosion and mitigate floods in low-lying areas, while coastal forests absorb energy from storm surges and act as a barrier for inland areas.

Forests provide tangible resources to humans, such as water, food, medicine, and timber. Forests provide direct and indirect services that support local and global industries, and their conservation and management are now mandated by various international treaties. Apparently, forests in developing and less-developed countries are more vulnerable due to population pressure and unsustainable utilization of the resources and services they provide.

Concept map activity as a tool for visualizing these relationships

Identifying the biotic and abiotic components of forest ecosystems and identifying their relationship is key to understanding the overall role of forests. Plants, animals, and microbes are interacting, keeping the dynamics complex and interdependent. For example, forests are recognized not only as sources of food, timber, and medicine but also for their role in climate regulation, which encompasses CO2 absorption, climate change mitigation, and microclimate regulation (i.e., temperature, rainfall, humidity). Similarly, understanding the forest's role in air and water purification emphasizes the plants' production of oxygen, absorption of pollutants, and prevention of soil erosion. Finally, forests are natural barriers to the impacts of natural disasters like flash floods, storm surges, and even wildfires, by absorbing excess water and slowing wind velocity

References

Pan, Y., Birdsey, R. A., Fang, J., Houghton, R., Kauppi, P. E., Kurz, W. A., Phillips, O. L., Shvidenko, A., Lewis, S. L., Canadell, J. G., Ciais, P., Jackson, R. B., Pacala, S. W., McGuire, A. D., Piao, S., Rautiainen, A., Sitch, S., & Hayes, D. (2011). *A large and persistent carbon sink in the world's forests. Science, 333*(6045), 988–993. https://doi.org/10.1126/science.1201609

Secretariat of the Convention on Biological Diversity. (2001). The value of forest ecosystems (CBD Technical Series No. 4). Secretariat of the Convention on Biological Diversity).

PART III

Essays on Invasive Plants by Region

About the Essays

As COIL is primarily student centered, I'd like to highlight the work of the students, many of whom are seeking degrees in the various areas of land management. The essays that follow represent the heart of the students' COIL experience. Working individually and in small groups, students brought their own regional knowledge and academic research to bear on a shared problem, producing work that is both rigorous and deeply rooted in place.

Glenville State University

Elijah Armes and Jessica Armstrong discuss "The Ecological Impacts of Invasive Asian Bittersweet." This species (*Celastrus orbiculatus*) grows around the stems of native plants, "smothering, strangling, and using all available resources." Rachel Ames and Cameron Beachler write about Autumn Olive (*Elaeagnus umbellata*), an East Asian shrub first introduced into the United States in the 1930s for erosion control but which spread so rapidly that it outcompeted native vegetation. Bryce Cogar informs readers about garlic mustard (*Alliaria petiolata*), originally cultivated for medicinal and culinary purposes, but now a pervasive species invading most of the northeastern and midwestern United States. Dawson Brown and David Booher discuss the history and characteristics of Japanese Barberry (*Berberis thunbergii*), a woody shrub that rapidly shades out native species.

Benjamin Gabehart and Veronica Foster-Toledano present their research on Japanese Honeysuckle (*Lonicera japonica*), first introduced as an ornamental plant used for erosion control, but now considered an aggressively invasive, smothering plant. Joshua Weister and Sam Turley discuss Japanese Stilt Grass (*Microstegium vimineum*), accidentally introduced to the United States as packing material, which is highly invasive, changing soil pH and reducing native tree seedling regeneration. Zachary Holstion and Jackson Mallett write about Multiflora Rose (*Rosa multiflora*), an Eastern Asian shrub introduced as rootstock for ornamental roses, as living fences, and for erosion control. Kieren Poff researches the ecological costs of Tree of Heaven (*Ailanthus altissima*), a highly invasive tree originally used for its medicinal and ornamental characteristics. Finally, Spencer Nolan discusses Kudzu (*Pueraria lobata*), a highly invasive vine with which I am personally familiar. I grew up in Southcentral Arkansas, where Kudzu vines were everywhere, taking over trees, fences, and telephone poles. Nolan notes how Kudzu is becoming a growing problem in West Virginia.

Southern Luzon State University

Students from Southern Luzon State University examined several invasive species with significant ecological and economic consequences in the Philippines and beyond. John Abejo, Alvin Andrade, and Mark Rodil present their research on the Giant Sensitive Plant (*Mimosa invisa*), an invasive shrub and climber that quickly produces dense thickets displacing native vegetation and degrading agricultural land. Edralin Angara, Bernadette David, and Blisslyn Nañola present their research on Cogon Grass (*Imperata cylindrica*), accidentally introduced through human interaction and now considered one of the world's ten worst weeds, forming dense monocultures highly resistant to eradication. Ronald Kristoffer Gayeta, Bernadette Mancenido, and Wilson Ramos investigate *Lantana camara*, a flowering shrub whose toxicity to livestock and ability to form impenetrable thickets make it one of the most damaging invasive plants in tropical regions worldwide.

Earl Joseph Tinamisan and Roan Gendrano examine Big-Leaf Mahogany (*Swietenia macrophylla*), a commercially valuable timber species whose aggressive natural regeneration outside its native range can suppress local forest biodiversity. Ronalyn Mañosca, Mary Anne Causapin, and Jaeshienne Therese Payumo report on Water Hyacinth (*Eichhornia crassipes*), a free-floating aquatic plant that forms dense mats blocking sunlight, depleting oxygen, and choking waterways to the detriment of aquatic life. Faith Ezaira Ponciano and Hazel Grace Yacon round out the Southern Luzon contributions with their study of Ipil-Ipil (*Leucaena leucocephala*), a fast-growing leguminous tree that, despite its uses in agroforestry, can escape cultivation and overrun native plant communities.

University of the Philippines Los Baños College of Forestry and Natural Resources

Researchers from the University of the Philippines Los Baños College of Forestry and Natural Resources bring additional depth to the study of aquatic and terrestrial invasives in the Philippine context. Elisha Ruth Sarita and Norgee Parreño revisit Water Hyacinth (*Pontederia crassipes*) with a focus on management solutions, noting that a single plant can produce thousands of seeds that remain viable in sediment for up to thirty years. Dailyn Rui Aguila and Russell Ivan Bautista trace the history and invasive spread of *Mikania micrantha*, a fast-climbing vine known as "mile-a-minute weed" that can grow several centimeters per day and smother entire crop fields and forest understories. Jayvee Velasco Mendoza examines *Thunbergia grandiflora*, a visually striking ornamental vine whose rapid vegetative growth allows it to blanket the forest canopy and eliminate light available to native species beneath.

Universiti Tun Hussein Onn Malaysia

Students from Universiti Tun Hussein Onn Malaysia contribute studies on invasive plants threatening Malaysian ecosystems, farms, and roadsides. Yung Li Zhi, Teu Yoke Yan, and Aqilah Awg Abdul Rahman profile Creeping Wood Sorrel (*Oxalis corniculata*), a low-growing groundcover whose seeds are dispersed explosively and whose tolerance of a wide range of soils and climates makes it notoriously difficult to control. Janice Lee Jing Yan, Tan Yan Zi, and Aqilah Awg Abdul Rahman examine Soap Bush (*Miconia crenata*), a shrub that colonizes disturbed sites rapidly and whose prolific seed production allows it to establish dense stands crowding out native vegetation. Muhammad Haziq Asyraaf Abu Bakar, Zahirulhaq Assiddiq Azmi, and Aqilah Awg Abdul Rahman study Billygoat Weed (*Ageratum conyzoides*), a fast-spreading annual weed that releases allelopathic chemicals into the soil, suppressing germination and growth of surrounding plant species.

Auni Syazana Zahar, Fatin Nurafiqah Mohd Fikri, Nur Faiqah Zani, and Aqilah Awg Abdul Rahman investigate Bermuda Grass (*Cynodon dactylon*), an aggressive turfgrass species that spreads through both seeds and underground rhizomes, enabling it to invade natural areas, roadsides, and cultivated fields. Finally, Nur Ayu Natasha Abdul Muizz, Nurzatul Syahirah Sunatwi, and Aqilah Awg Abdul Rahman present their research on Touch-Me-Not (*Mimosa pudica*), a spiny, fast-spreading weed whose rapid growth and seed longevity allow it to persist in disturbed soils despite control efforts.

These students' essays offer more than an inventory of troublesome plants; they represent a genuinely international conversation about one of the most pressing challenges in land management today. Invasive species recognize no borders, and neither does the knowledge needed to address them. Readers will find that the research produced by students in West Virginia, the Philippines, and Malaysia speaks to shared problems even as it reflects the distinct ecosystems and land-use traditions of each place. That combination of local expertise and global perspective is precisely what makes COIL so valuable.

A

Glenville State University

Ecological Impacts of Invasive Asian Bittersweet (*Celastrus orbiculatus*)

Elijah Armes & Jessica Armstrong

Introduction

Celastrus orbiculatus, or commonly referred to Asian bittersweet, is an invasive species found in Eastern North America as a profound invasive species. Asian bittersweet is commonly referred to as a variety of names such as Ornamental bittersweet, Asiatic bittersweet, and Chinese bittersweet. The plant is identifiable as a woody vine that grows around the stems of native species of plants. The morphology of Asian Bittersweet is described in the article "MN NWAC Risk Assessment Worksheet" as it describes it as "leaves are simple, variable, but typically obovate (egg-shaped, broadest toward the tip) to orbicular (rounded), alternately arranged, toothed, and glabrous with distinctly pointed tips" (Minnesota Department of Agriculture & Minnesota Nursery and Landscape Association, 2016). There is a native bittersweet found within North America called *Celastrus scandens*, which the Asian variant can quickly hybridize with.

Fig. 1 The Asian Bittersweet vine tightly wrapped around a tree, allowing for rapid vertical growth, found on the United States Department of Agriculture website (Miller, n.d.)

Native Range and Method of Introduction

Asian Bittersweet is native to several Asian countries such as Japan, China, and Korea. In these countries Asian Bittersweet can be found to have an important role in ecosystems. The species is also found to be well adapted to a range of environmental conditions, the ability to grow in a variety of environments has proven to be detrimental to many plant species along the Appalachian Mountains (Invasive species in Connecticut, n.d.) It is also to be considered as a fire tolerant species, which means it has adapted to be able to withstand fires and continue to grow afterwards. With the ability to survive in a range of environmental condition, it has created many issues across the North East United States.

Introduction Method

Asian bittersweet (*Celastrus orbiculatus*) was imported into the United States around the year of 1879, originally brought over as an ornamental plant and to be used for erosion control. The bright red berries and twirling patterns continue to contribute to the ornamental scene. Since then, it has been reported to have been established in at least 33 states and labeled as invasive in 21 states (Wooten, 2013).

Fig. 2 The berries of Asian Bittersweet, which stay on the plant from summer into winter, found on the United States Department of Agriculture Website (Swearingen, n.d.)

Fig. 3 Overview of how Asian Bittersweet can take over native species found within the field (Armes, 2025)

Ecological and Economic Impacts

Oriental bittersweet can be absolutely devastating to environments in the eastern United States. The lack of large amounts of biological control and natural predators allow the vine to spread rapidly over the landscape. It has the ability to adapt to a large range of conditions and can easily become established in disturbed habitats, fields and pastures, roadsides, and even in closed canopy forests (Wooten, 2013). Its vigorous root sprouting and suckering capabilities allow it to quickly colonize any area it enters. The vines spread quickly and cover all vegetation in its path eventually out competing the native plants by smothering, strangling, and using all the resources available (Oriental Bittersweet, n.d). The presence of Oriental bittersweet has also been tied to altered pH levels, calcium, potassium, and magnesium levels. It increases nitrogen mineralization and litter decomposition. (Wooten, 2013).

It continues to wreak havoc by hybridizing with the native bittersweet (*Celastrus scandens*). This hybrid is considered to be just as much, if not more, of a threat to native vegetation (Lynch, 2009). All these traits together cause incredible amounts of damage and add to the $1.288 trillion worth of damage from invasive species over the past 50 years. (USDA, 2025)

Control and Eradication Methods

Once it was determined to be a problem invasive, Asian bittersweet quickly became one of many target species to kill and remove from the landscape. But due to its very competitive traits, such as shade tolerance and prolific seed production, control and eradication have been no easy feat. Even with the extensive knowledge gained on this species, the so-called perfect eradication method is still unknown. However, the potential methods that have been devised are described on the next page.

Mechanical

Mechanical control includes any methods that use any form of equipment (i.e. tractor, excavator, dozer, etc.), tools, or simply hand pulling. Asian bittersweet is a prolific root-sprouter and stump-sprouter, so mechanical control leaves a great deal of variability in its success of controlling and removing the invasive. Attempting to remove the plant mechanically but being unsuccessful in removing the entire root system allows the plant to respond to the damage by resprouting. There have been studies conducted that show mechanical removal as slightly effective, but higher mortality rates are apparent in combination treatments where mechanical removal is followed up by a chemical treatment. Prescribed burning has also been used to attempt eradication but has usually proven unsuccessful by merely killing the stems but not the root system. This allows for resprouting from the stump and the roots.

Chemical

There have long been numerous types of herbicides on the market. Ranging from species specific, non-specific, oil carriers, pellets, etc., the user can find the seemingly perfect herbicide for the task. In common place, Glyphosate has been used for many decades as a non-selective, post-emergent herbicide, that directly enters the plant, and it transmitted throughout the plant system (Wooten, 2013). In general, Glyphosate is applied as a foliar spray directly onto the plant's foliage, or as a "cut stump" treatment directly to freshly cut stems and stumps. Cut stump treatments are most used for plants that have root sprouting and stump sprouting reproductive characteristics. From a study completed at the Presque Isle State Park in Erie, Pennsylvania, "The most effective treatment to suppress *C. orbiculatus* is by preforming window cuts on all stems followed by a foliar herbicide treatment of 3% triclopyr and 6% glyphosate five weeks post cut" (Wooten, 2013). Triclopyr is another common herbicide to use as it is non-toxic to most animals and insect species, has no effect on conifers and grasses, but still controls broadleaf weeds and other woody plants. (Wooten, 2013. Oriental Bittersweet n.d) Yet another study showed that chemical treatments yielded only semi-positive results, meaning that there was an initial decrease in non-native presence but did not remain suppressed. There were better results in the plots that were mechanically, or mechanical/chemical combination control (Farmer, Ward, Horton, Clarke 2016)

Biological

All vegetation, native and non-native, are always at risk of harm conducted by some biological means. Whether it's an insect, fungus, disease, wildlife, etc. Where non-native species are out of their native range, often times they have no significant biological control. One study, conducted in Winona, Minnesota, showed that invasive oriental bittersweet is impacted by White-Tailed deer browsing. The vine was heavily browsed in the winter months, resulting in 99% of all terminal shoots and 70% of all lateral shoots of the study plants being browsed. (Mundahl, Borsari, 2016.) However, plants that have previously been treated with herbicides or other chemicals are likely to be avoided. Unfortunately, no other notable biological control that has shown impeccable abilities of control and/or eradication has been identified and put into practice.

Oriental Bittersweet (*Celastrus Orbiculatus*) Concept Map

The following figure is a concept map devised to show a summary of the relationships between Oriental bittersweet and its non-native habitat:

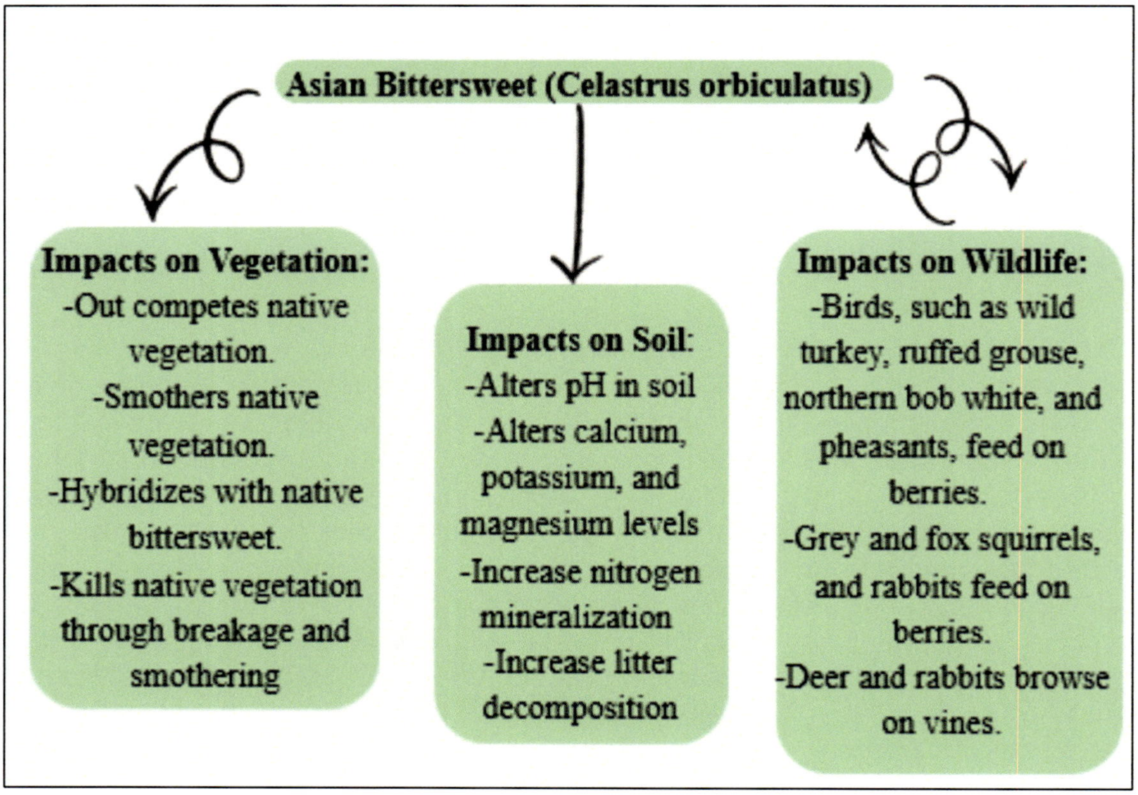

Fig. 4 Concept Map of Oriental Bittersweet (Armstrong, 2025).

The concept map is a brief diagram of the interaction between Asian bittersweet and factors of the environment. The impacts on vegetation include out-competing native vegetation, hybridizing, smothering and strangling native vegetation, etc. Asian bittersweet alters the soil by increasing nitrogen mineralization, increasing litter decomposition, and altering pH, calcium, potassium, and magnesium levels. It acts as a food source for birds, grey and fox squirrels, deer, and rabbits, and these species in turn spread the seeds aiding in the reproduction. However, this only furthers the rapid destruction.

References

Armes, E. (2025). *Oriental bittersweet in the Field* [Photograph].

Armstrong, J. A. (2025). *Celastrus orbiculatus Concept Map.*

Economic and social impacts. Economic and Social Impacts | National Invasive Species Information Center. (n.d.). https://www.invasivespeciesinfo.gov/subject/economic-and-social-impacts

Farmer, S., Ward, J. R., Horton, J. L., & Clark , H. D. (2016, October 10). Southern Appalachian urban forest response to three invasive plant removal treatments. Ashville; Biology Department, University of North Carolina Asheville.

Invasive species in Connecticut. (n.d.). Town of Griswold Wetland & Conservation. https://cms8.revize.com/revize/griswoldct/Document%20center/Wetland%20&%20Conservation/new-website—invasive-species-in-ct.pdf

Kurtz, Cassandra M.; Hansen, Mark H. 2018. An assessment of oriental bittersweet in northern U.S. forests. Res. Note NRS-251. Newtown Square, PA: U.S. Department of Agriculture, Forest Service, Northern Research Station.

Minnesota Department of Agriculture & Minnesota Nursery and Landscape Association. (2016, September 29). *Oriental bittersweet (Celastrus orbiculatus) – Risk assessment worksheet* [PDF]. https://static1.squarespace.com/static/57539006044262fce01261c5/t/5a8ee5c1085229db68d4b1e9/1519314371071/OrientalBittersweet_MDARA000010B_09_29_2016.pdf

Lynch, A. L. (2009). Investigating Distribution and Treatments for Effective Mechanical and Herbicide Application for Controlling Oriental Bittersweet (*Celastrus orbiculatus Thunb.*) Vines in an Appalachian Hardwood Forest. Michigan; Proquest LLC.

Miller, J. H. (n.d.). *Round Leaf Bittersweet*. U.S. Department of Agriculture. https://www.fs.usda.gov/database/feis/plants/vine/celorb/all.html

Nowak, C. A., & Peck, C. J. (2016, July). Large Oriental Bittersweet Vines Can Be Killed by Cutting Alone: Implications for Utility Arboriculture and Other Hazard Tree Work. Oriental Bittersweet - Celastrus orbiculatus. (n.d.).

Swearingen, J. M. (n.d.). *Oriental Bittersweet*. U.S. Department of Agriculture. https://www.fs.usda.gov/database/feis/plants/vine/celorb/all.html

Wooten, J. (2013, May 13). Techniques to Suppress Invasive Oriental Bittersweet (*Celastrus orbiculatus*) on Presque Island State Park in Erie, Pennsylvania.

Ecological Effects of Autumn Olive (*Elaeagnus umbellata*)

Rachel Ames & Cameron Beachler

Abstract

Autumn olive (*Elaeagnus umbellata*) a shrub from East Asia, introduced in to the United States in the 1830s for erosion control and for habitat and food for wildlife. Its aggressive growth, high seed production, and the nitrogen fixation abilities has allowed it to spread rapidly and outcompete native vegetation. This paper examines the species history, identifying characteristics, ecological and economic impacts, influences on vegetation structure, biodiversity, and soil chemistry. Observations and existing research highlights on how dense thickets are able to disrupt native regeneration, alter habitat, change the dynamics in nutrients in soils in reclaimed and natural areas. Management strategies have been done through mechanical and chemical controls, education of the public and landowners have also been used.

Introduction

Autumn olive also known as by its scientific name *Elaeagnus umbellata*. This species originates in East Asia and was introduced in the United States in the 1830s, planted for control of erosion, habitat improvement for wildlife, and as an ornamental. Autumn Olive has the ability to fix nitrogen and grow in poor soils. Over time, autumn olive has started to become an issue, its aggressive growth and high seed production allows for the species to outcompete native species and vegetation and is also able to alter soil chemistry. Autumn olive is threatening native communities and reducing biodiversity in an ecosystem. To prevent further ecological impacts, an understanding of disturbances and management practices is needed to protect native habitats against the spread of the invasive species.

Species History

Autumn olive was first introduced in the 1830s as an ornamental plant. It also was introduced because it could provide habitat and food for wildlife species. It was planted by the Soil Conservation Service to help stop erosion on roadsides and ridges. In the 1950s it was planted as well to help disturbed sites such as mining sites. Now after almost 200 years, the autumn olive has become a major problem and is hurting native trees and plant species.

Characteristic, and Information

Autumn olive is a shrubby tree that can grow up to 20 feet tall and 30 feet wide. It has elliptically shaped leaves and silver scales on the backside of the leaf. The autumn olive's flowers are bell shaped and are cream to pale yellow in color. The berries it produces are red and dotted with silver scales.

Native Range and Method of Introduction

Autumn olive is native to most parts of Asia and like previously mentioned, the Soil Conservation Service planted it for erosion control and wildlife values.

Ecological and Economic Impacts

Autumn olive has several ecological impacts, including its ability to fix nitrogen, alter vegetation structure, change species composition and affect habitat quality. An effective trait to autumn olive is its production of seeds and the dispersal through birds and animals. According to the United States Department of Agriculture Forest Service (USDA), a single autumn olive fruit contains a see that can readily germinate and establish. According to the U.S. Forest Service, out of 24 surveyed northern states, 19 have documented the presence of autumn olive, with an estimated occurrence of 6.7% (Kurtz & Hansen, 2016; Michigan Department of Natural Resources & Michigan Natural Features Inventory, 2012). Autumn olive is able to thrive in low fertile soils, it has nitrogen-fixing abilities meaning that it is able to convert atmospheric nitrogen to forms that plants can use, such as ammonia and other nitrogen compounds. The ability that autumn olive has can be very useful in reclamation of land but leads to consequences to an unintended ecosystem. One example would be the presence of the species on coal mines in Appalachia. This had increased nitrogen levels in the soil, but does reduces potential establishment of native species (Franke, Zipper, & Barney, 2018) With the change in nitrogen the plant composition shifted and reduced suitability in habitats for species adapted to soils poor in nutrients.

Vegetation Structure, competition and Successional Disruption

Autumn olive establishes in disturbed, open, and or edge environments. This invasive often forms multiple stems that form into dense thickets. Thickets formation have the potential to create a change in the dynamics of sites it invades in a physical or biological dynamic. Autumn olive intercepts light and creates shading, when it reaches a mature height its canopy can create a shade for understory and the ground surface, meaning it can reduce light availability in growth and germination for native woody and herbaceous species.

Autumn Olive is able to form thickets that are able to harm or suppress native regeneration of woody or herbaceous plants. This is important because native plants need space, sunlight and nutrients in the soil, when Autumn Olive forms it thickets it reduces the chances for native plants to establish growth (Kurtz & Hansen, 2016).

Autumn Olive can become a ground cover dominant and space competition, being multi-stemmed and having a lateral spread it creates a limited space available on the soil surface and creates a physical barrier for native seedlings. Having a dense root crown system with a dense branching system it keeps other species from being able to establish growth.

In disturbed sites Autumn Olive can negatively impact slower-growing native species, particularly in restoration areas that rely on establishment of native shrubs and trees. With mature Autumn Olive thickets, this can disrupt normal successional progress. This shows that Autumn Olive can create structural changes through both physical and ecological processes.

Successional Trajectory Alteration

With the natural or reclaimed lands, succession in vegetation involves the colonization of herbaceous species, such as shrubs and woody trees. The aggressiveness and early establishments in open sites cause a disruption in normal succession because it establishes a shrub layer that is persistent that can alter or prevent the growth to a tree dominant system. With its nitrogen fixing ability, Autumn Olive can alter soil fertility by increasing nitrogen in poor soils. This may favor fast growing shrubs and opportunistic species over slow growing tree seedlings that are adapted to nutrient-poor soils. In reclaimed mine cases, the survival of native tree seedlings significantly dropped where Autumn Olive was present. With the removal treatments, they saw that establishment improved over the period of two growing seasons, suggesting that Autumn Olive presence can prevent or delay desired successional growth towards a hardwood cover (Franke, Zipper, & Barney, 2018).

Impacts: Biodiversity, Habitat, Soil, and Hydrology

While Autumn Olive is tied to vegetation competition and structure, it affects and creates dominance in habitats, biodiversity, soil, and hydrology. Shifting into dense monocultures creates few niches for native plants, birds, mammals, and insects relying on variable understory conditions. A study found that the branch morphology found in denser branching structure of Autumn Olive did not affect the perching behavior significantly in Midwest habitats for birds. This shows that some species can use this invasive for nesting sites or temporary perching sites, but this is a short-term interaction. Long terms interactions are seen to be more complex.

As the invasive spreads, it is able to reduce native plant diversity and change vegetation structure. Native plants that are important for wildlife are hindered by Autumn Olive, it is able to decrease the amount of browses available for wildlife. This can also change composition of small mammals and insect communities because many depend on ground cover mixed with wildflowers, native grasses, and shrubs. The leaf structure form of Autumn Olive has very few nesting opportunities in microhabitats and provide less food availability.

Nutrient changes beneath Autumn Olive stands can lead to a larger leaching of nitrates, and have a downstream impact on sensitive landscape water quality. Soil alteration and rapid root growth can influence soil moisture regimes.

Economic and Land Management

When Autumn Olive establishes on a site the removal and control can take multiple years. Treatments can often require three to five years of following treatments, the repeated treatments, monitoring, equipment, chemical, native species re-establishment and restoration costs could be later needed (Chandran & Loyd, 2015).

Control and Eradication Measures

Autumn Olive management requires a mechanical, chemical, and biological. Each method has limitations and advantages when used appropriately. In mechanical control it consists of mowing, cutting, uprooting, physical removing, or burning Autumn Olive.

Hand Pulling

Pulling by hand is effective for small seedlings when the root system is more easily extracted. This method should be completed in the early spring when the invasive has not been able to move nutrients like stored carbohydrates to the above ground tissues, the roots are smaller and the invasive is less resilient (Czarapata, 2005). Hand pulling is labor intensive and is usually only seen in small scale sites and natural communities. Removal of the complete root system is critical because any remaining fragments can resprout. This method minimizes disturbances of soils and herbicide use.

Cutting and Mowing

In the sense of larger shrubs mechanical cutting, brush mowing, and chainsaw cutting treatments can be used. This temporarily reduces above ground mass, however Autumn Olive promotes resprouting from the crown and stumps. This can make mechanical treatments alone not useful (Integrated Pest Management). A recut at regular intervals can be done to deplete the reserves of carbohydrates in roots, limiting vegetative spread. This method is most useful in field that have been abandoned, reclaimed mine sites, and forest edges were chemical application is restricted when near sensitive soils or near native vegetation. With this a cut-stump method can be completed with the immediate herbicide application.

Prescribed Fire

Burning can be useful in fire adapted areas. It can reduce the population of seedlings and small shrubs when applied during the growing season. Autumn Olive, however, is resilient to burning, mature plants are hardy killed with fire because of the deep root system and protected root crowns. Generally fire is recommended as a supplemental treatment within a management plan. Often seen combined with chemical or mechanical treatments controlling reinvasion and resprouts (Stark, 2001).

Chemical Control

Autumn Olive can be controlled using chemicals such as herbicides. For the application of herbicides and to minimize non targets the most effective time to do the application is most effective in the late summer to early fall where the plant is in full leaf and prior to drop. A cut sump or basal treatment it to be used more towards the dormant seasons. When looking at doing an herbicide application following the label and local standards are important. We use as an active ingredient the use of Glyphosate and triclopyr are the main active ingredients. Glyphosate is the most effective in the cut-stump application and so is a direct foliar spray for fresh cut surfaces (WVU Extension). Triclopyr is most effective in cut-stump, basal bark, and foliar using the appropriate labeled concentrations.

Application for the cut-stump method is when the stems are cut near or at ground level then are immediately treated on the cambial ring and surface. Basal bark application is done at lower volume applications applied to stems that are thirty five to forty centimeters or thirteen to fifteen inches in diameter. Foliar is best for dense patches of sampling Autumn Olives (WVU Extension, Missouri Department of Conservation, Michigan Natural Features Inventory).

Field Photos

Fig. 1 A dense patch of Autumn Olive growing alongside an edge habitat

Fig. 2 A closer view of a dense patch of Autumn Olive show the shrubs dominate the understory, forming arching branches that block sunlight from reaching the forest floor.

Fig. 3 A young Autumn Olive growing near the base of a hickory tree shows that the leaves are arranged alternately along a slender stem and are oval in shape.

Species Concept Map

AUTUMN OLIVE

Cause of Spread

- Bird & wildlife spread seeds
- Human planting for erosion control/habitat creation
- Growing aggressively by reproduction

Evidence (Impacts We See)

- Crowds out native plants
- Reduces biodiversity
- Nitrogen fixation alters soil chemistry
- Takes over old fields & forest edges

Solutions

Further Research & Monitoring	Chemical Treatments (herbicide application)	Public Education Landowner Outreach

This concept map outlines the major factors that contribute to the spread of Autumn Olive. The map above starts the causes of spread highlighting the main ways the invasive species spreads across landscapes. The spread is primarily caused by birds, human planting, and its ability to grow aggressively. The next arrow leads to the evidence (impacts we see). This lists some of the ecological problems that Autumn Olive causes. This includes crowding out native species, reducing biodiversity, altering soil chemistry through nitrogen fixation, and taking over fields and forest edges. The final section is the solutions that shows the three primary management solutions. This includes further research and monitoring, chemical treatments such as herbicide application, public education and landowner outreach.

Conclusion

The history of Autumn Olive in the U.S. reflects how the introduction can cause lasting ecological challenges. Originally planned for wildlife habitat, reclamation, and erosion control. Autumn Olive have shown the ability to spread quickly, alter soil chemistry, and can form dense thickets outcompeting natural native vegetation. With having nitrogen fixation abilities, aggressive reproduction, and its tolerance of poor soils allowing it to have the ability to change plant communities, disrupt successional pathways, and reduce biodiversity in forests, open fields, and reclaimed lands. Many different methods have been explained to treat the issue with the spread of Autumn Olive, from hand pulling, to chemical treatments. Since being left unchecked the species has spread rapidly and uncontrolled and now efforts are being done to control the spread of Autumn Olive.

References

Autumn olive: Your invasive, seedy neighbor. The Nature Conservancy. (n.d.). https://www.nature.org/en-us/about-us/where-we-work/united-states/indiana/stories-in-indiana/autumn-olive/

Chandran, R. & Loyd, B. (2015). *Autumn Olive* (ANR-IPM-15-012) West Virginia University Extension Service https://extension.wvu.edu/files/d/d575d145-38f0-4927-8262-afe6faa136e1/autumn-olive-formatted.pdf?

Franke, M. E., Zipper, C. E., & Barney, J. N. (2018). *Native hardwood tree seedling establishment following invasive autumn-olive (Elaeagnus umbellata) removal on a reclaimed coal mine.* Invasive Plant Science and Management, 11(3), 155-161. https://doi.org/10.1017/inp.2018.19

Kurts, C. M., & Hansen, M. H. (2016) *An Assessment of Autumn Olive in Northern U.S. Forests*. Research Not NRS-204. U.S. Department of Agriculture, Forest Service, Northern Research Station https://www.fs.usda.gov/nrs/pubs/rn/rn_nrs204.pdf

Michigan Department of Natural Resources & Michigan Natural Features Inventory. (2012, February). *Invasive Species – Best Control Practices: Autumn Olive (Elaeagnus umbellata).* https://mnfi.anr.msu.edu/invasive-species/AutumnOliveBCP.pdf

Michigan Natural Features Inventory (2012). *Best Control practices for autumn olive (Elaeagnus umbellata Thunb.)* Michigan State University Extension. https://mnfi.anr.msu.edu/invasive-species/AutumnOliveBCP.pdf

Missouri Department of Conservation. *Autumn Olive control*. Missouri Department of Conservation. https://mdc.mo.gov/trees-plants/invasive-plants/autumn-olive-control

Seeley, N. S. (2022). *Invasive autumn olive's impacts on native red oak growth rate and germination.* (Undergraduate independent studies thesis, The College of Wooster). Department of Biology, The college of Wooster. https://wooster.edu/2022/04/26/nathaniel-seeley/

Stark, C. (2001). *Control of Elaeagnus Umbellata (Autumn Olive).* University of Minnesota, Department of Forest Resources. https://conservancy.umn.edu/server/api/core/bitstreams/85274c22-c918-4391-91e8-e8169dfebb32/content

Wallace, V., & Siegel-Miles, A. (2022). *Invasive plant factsheet: Autumn Olive (Elaeagnus Umbellata).* UConn Extension https://ipm.cahnr.uconn.edu/invasive_plants_autumn_olive/?utm

West Virginia University Extension Service (2015). *Autumn olive (Elaeagnus umbellata): Invasive species fact sheet.* West Virginia University. https://extension.wvu.edu/files/d/d575d145-38f0-4927-8262-afe6faa136e1/autumn-olive-formatted.pdf

Forest Impacts of Garlic Mustard (*Alliaria petiolate*)

Bryce M. Cogar

History and information of Garlic Mustard

Garlic Mustard (*Alliaria petiolate*), also known as Garlic Root, is a biennial herb belonging to the mustard family (Brassicaceae). The plant's life cycle spans two years. In the first year, called the rosette stage, it forms a low-growing rosette of kidney-shaped leaves that stay green throughout the winter. In the second year, the flowering stage, it develops one or more flowering stalks that can reach up to 3 feet (1 meter) high. The leaves on these stalks are alternate, triangular to heart-shaped. It flowers early in the spring (April-May), producing clusters of small, white flowers with four petals. Before the plant dies, it produces long, narrow seed pods called siliques, which contain small, black seeds. A key identifying characteristic is all parts of the plant smell like garlic when crushed. Garlic Mustard is highly shade-tolerant and thrives in habitats such as the understory of deciduous forests, floodplains, roadsides, and disturbed areas.

Native Range and Method of Introduction

The native range of Garlic Mustard is Europe and parts of Asia. Its introduction to North America dates back to 1868, when it was first reported on Long Island, New York. It is generally believed that early European settlers intentionally introduced it for medicinal and culinary purposes, valuing it as an edible herb. After its introduction, Garlic Mustard escaped cultivation and began spreading aggressively. It is now a pervasive invasive species, particularly throughout the northeastern and midwestern United States, including states like West Virginia.

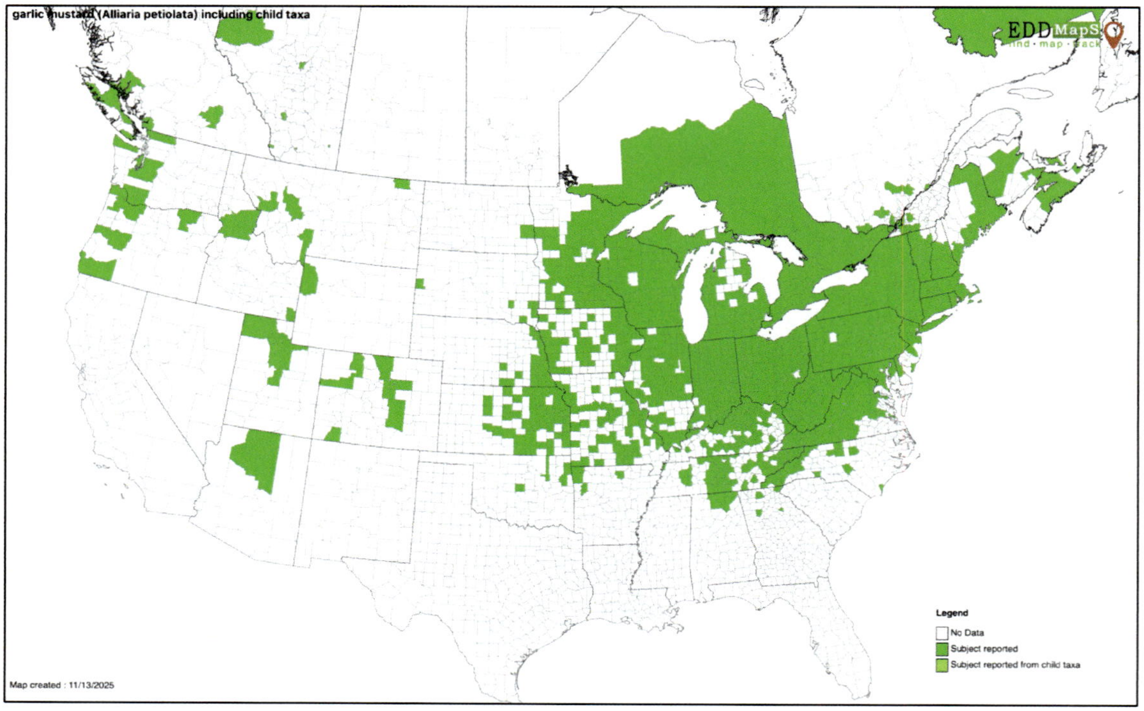

Fig. 1 *Alliaria petiolata* distribution by county, 2025, from eddmaps.org

Ecological and Economic Impacts

Garlic Mustard poses a severe threat to the native biodiversity and hardwood forests of West Virginia. Its most significant ecological impact is through competition and displacement of native species. The plant forms dense, monotypic stands that aggressively out-compete and shade out native spring ephemeral wildflowers, such as trilliums, spring beauty, and toothworts, by having a longer growing season, starting earlier in the spring and staying green later in the fall. Furthermore, Garlic Mustard exhibits allelopathy; it releases toxic chemical compounds (allelochemicals) into the soil. These compounds inhibit the growth of competing native plants and are known to interfere with the symbiotic relationship between mycorrhizal fungi and native trees (like oaks), which can suppress the growth and survival of native tree seedlings. A particularly devastating impact is its role as a "toxic decoy" for the West Virginia White butterfly (*Pieris virginiensis*). Because Garlic Mustard belongs to the same plant family (Brassicaceae) as the butterfly's native host plants (toothworts), the butterfly is attracted to it and lays its eggs on it. However, the larvae cannot survive on Garlic Mustard, essentially making it a "toxic trap" that contributes significantly to the butterfly's decline and extirpation in certain areas.

The primary economic impact of Garlic Mustard is related to the substantial cost and long-term effort required for its control and management. This management is necessary across various land types, including public forests, natural areas, and private property. The financial burden includes the recurring expense of labor, purchasing herbicides, and the subsequent restoration efforts needed to re-establish native plant communities once the invasive species has been removed.

Control of Garlic Mustard

Effective control of Garlic Mustard demands a long-term commitment, often spanning 5–10 years, primarily due to its persistent seed bank, where seeds can remain viable in the soil for at least 5 years.

- **Mechanical Control:**
  - **Hand-pulling:** This is the most effective method for small infestations or newly invaded areas. Plants should be pulled in the early spring when the soil is moist, before seeds are set, and ideally, the entire root must be removed to prevent resprouting from the root crown.
  - **Cutting/Mowing:** Flowering stalks can be cut at ground level after the stalks elongate but before seeds form (typically late April to May). This can kill the plant or prevent seed production, but a follow-up is necessary as plants may re-sprout.
  - **Disposal:** All pulled or cut plant material, especially if it has flowers or seeds, must be bagged and disposed of in the trash (not composted) to prevent seed dispersal.
- **Biological Control:**
  - There are no widely available and approved biological control agents in the U.S. at this time. Research is ongoing for several species of weevils from its native range, particularly the root-mining weevil (*Ceutorhynchus scrobicollis*) and the seed-feeding weevil (*C. constrictus*), which have shown promise in trials.
- **Chemical Control:**
  - **Herbicide Application:** Non-selective herbicides like Glyphosate "Roundup" are effective but must be used with extreme caution to avoid harming native plants.
  - **Timing:** Application is most effective as a foliar spray during the early spring (March-April) or late fall when garlic mustard rosettes are actively growing, but most native plants are dormant, minimizing non-target damage. Triclopyr is another option, which is more selective for broadleaf plants and less harmful to grasses/sedges.

Fig. 2 from Garlic mustard pulling info: Core arboretum: West Virginia University

Fig. 3 Garlic mustard (*alliaria petiolata* Bieb.) (Cavara & Grande,) 2008)

Concept Map

The concept map appears on the next page. Garlic Mustard (*Alliaria petiolata*) significantly alters the forest ecosystem, primarily by acting as a strong ecological push factor. A major characteristic contributing to its dominance is its ability to emerge early in the spring, giving it a competitive advantage over native flora. It quickly out-competes native plants by forming dense, single-species stands, or monocultures. This extensive growth severely reduces biodiversity and alters the forest floor. The invasive plant also changes the soil environment through allelopathy, releasing chemical compounds that suppress the growth of other species. This invasive success affects various native animal and insect communities. For herbivores, Garlic Mustard often serves as an undesirable browse. Additionally, its presence negatively impacts pollinators and affects the reproductive success of native insect species that mistake it for their native host plants.

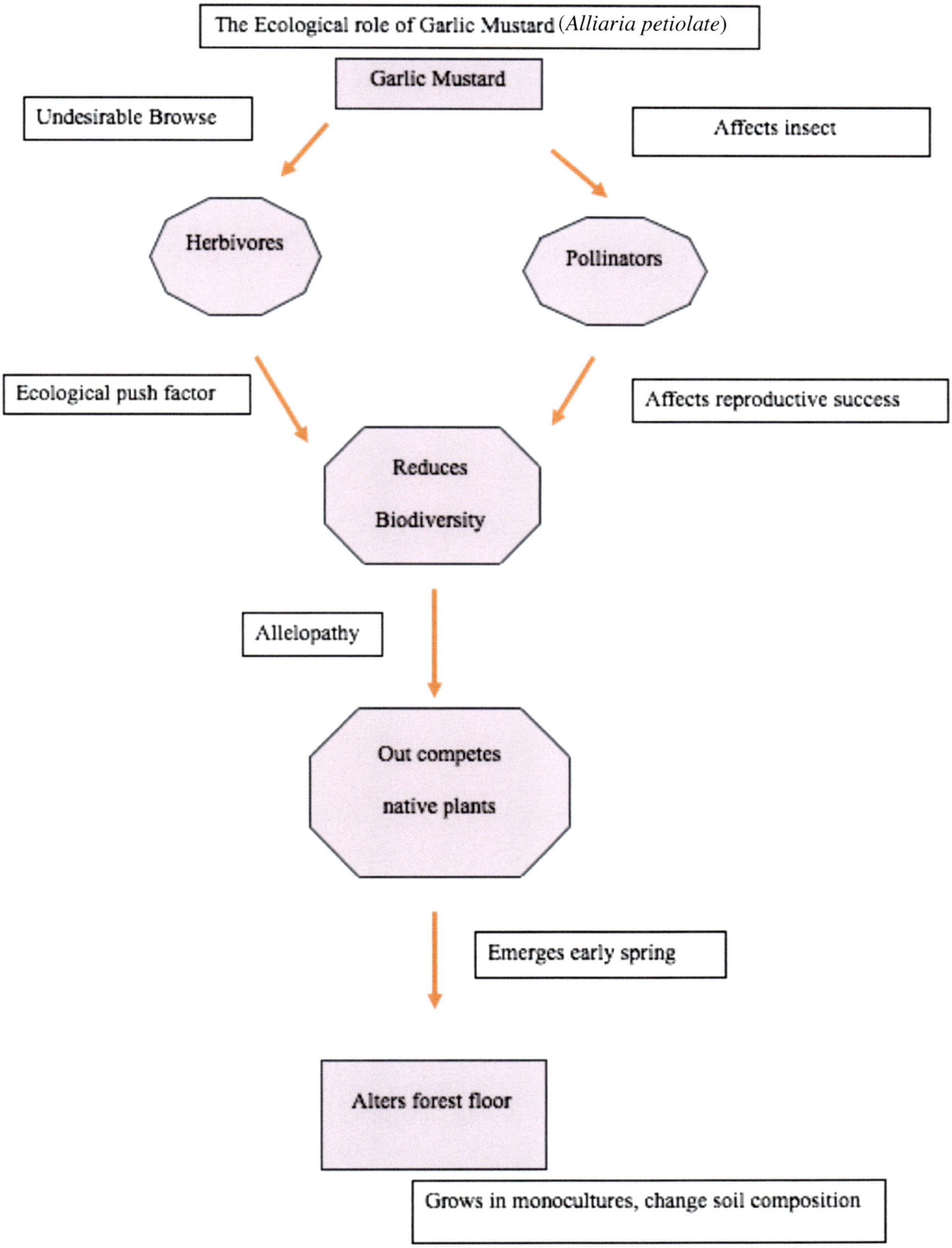

Fig. 4 Concept map for Garlic Mustard (*alliaria petiolate*)

References

Barker , J. (2018). Effects of white-tailed deer and invasive garlic mustard on native tree seedlings in an urban forest. *McMaster University* . https://macsphere.mcmaster.ca/handle/11375/24171

Garlic mustard (alliaria petiolata (Bieb.) Cavara & Grande) - eddmaps distribution - eddmaps. EDDMapS.org. (n.d.). https://www.eddmaps.org/distribution/uscounty.cfm?sub=3005

Garlic mustard (alliaria petiolata (Bieb.) Cavara & Grande). Invasive.Org. (2008, November 10). https://www.invasive.org/browse/detail.cfm?imgnum=2146038#

Garlic mustard pulling info: Core arboretum: West Virginia University. garlic mustard pulling info | Core Arboretum | West Virginia University. (n.d.). https://arboretum.wvu.edu/garlic-mustard-pulling-info

Garlic mustard. Garlic Mustard | National Invasive Species Information Center. (n.d.). https://www.invasivespeciesinfo.gov/terrestrial/plants/garlic-mustard

Garlic mustard: Invasive, destructive, edible. The Nature Conservancy. (n.d.). https://www.nature.org/en-us/about-us/where-we-work/united-states/indiana/stories-in-indiana/garlic-mustard/

Gazal, R. (2025) *Lecture slides 5-18* [PowerPoint Slides] Brightspace

Hollis, L. (2023, March 9). *Garlic mustard threatens the rare West Virginia white butterfly*. Invasive Species Blog. https://blog.invasive-species.org/2023/03/09/garlic-mustard-threatens-the-rare-west-virginia-white-butterfly/

Invasives in your woodland: Garlic Mustard (updated 2025). University of Maryland Extension. (n.d.). https://extension.umd.edu/resource/invasives-your-woodland-garlic-mustard-updated-2025/

Management and Control of Japanese Barberry (*Berberis thunbergii*)

Dawson Brown & David Booher

Species History

Japanese Barberry (*Berberis thunbergii*) originated in Japan. Japanese barberry was introduced to the United States as an ornamental plant in 1875. Seeds of Japanese Barberry were sent from a Russia supplier to a man in Boston Massachusetts named Arnold Arboretom in 1896. It was then planted at the New York Botanic Garden. It was at one point promoted as a substitute for Common Barberry (*Berberis vulgaris*), an exotic plant introduced and used by early settlers from Europe for organic fencing, dyes and canned foods, and later found to be a host for the black stem rust of wheat. Since then, it has grown immensely and proved to be an invasive species. (USDA) (Invasive.org)

Species Information

Japanese barberry occurs mainly in human-disturbed river and stream floodplains, forest edges, woods, meadows and fields. It is native to Japan and eastern China, as well as southwestern Russia. Invasive in the following states: Maine, New Hampshire, Vermont, New York, Connecticut, Massachusetts, Rode Island, Pennsylvania, Delaware, West Virgina, Virginia, Maryland, North Carolina, South Carolina, Georgia, Ohio, Tennessee, Kentucky, Indiana, Illinois, Michigan, Wisconsin, Missouri, Arkansas, Oklahoma, Iowa, Minnesota, South Dakota, Nebraska, Kansas, Colorado, and Washington state.

Species Characteristics

Japanese barberry is a woody shrub that can grow from four to six feet in height. It is tolerant of both sunlight and shade, making it very adept in a wide variety of habitats and light conditions. It grows well in full sunlight to deep shade and forms dense stands in closed canopy forests, open woodlands, wetlands, field edges and such locations. In turn it will grow rapidly shading out other species and then take over. Its invasive range consists most of the northeastern United States. It stretches from Maine over to Missouri and Minnesota and down to West Virginia and beyond. (USDA)

It is considered a woody shrub, due to its wide bushy crown and short woody stem. The bark of a mature Japanese barberry is ridged or plated and easy to peel off. It reproduces both asexually and sexually, spreading via seeds and rhizomes. When this species flowers it produces small yellow flowers that drop from the branches of this species. It has simple smooth margin leaves that occur in a cluster at each node along the branches of the plant. Its leaves are green during the growing season but, will turn a purplish red or maroon color in the fall. These leaves, like all deciduous plants, drop off in the late fall to early winter, leaving the reddish-brown bark visible. The inner bark has a mustard yellow color. Like most if not all barberry's the Japanese barberry has thorns along its branches. Unlike Common and Alleghany barberry the thorns of the Japanese barberry are "simple" and smooth. These thorns appear at the base of each node right below the cluster of leaves. These berries will show up in the summer and will hang on through the winter in either a single berry or in clusters of up to 4. The fruit of this species is very similar to that of the Common barberry, being bright red, fleshy, and hanging down from the branches.

Ecological Impact

In locations where it is well established, Japanese barberry displaces many native herbaceous and woody plants. In large infestations, its leaf litter and root system cause changes in the chemistry of the soil, making it more basic. Which in turn makes the site more favorable for the colonization of more Japanese barberry plants. Japanese barberry has also been shown to alter forest soil cycling through its relationship with invasive earthworms. Birds, particularly wild turkeys and other game birds consume the red berries and spread the seeds of Japanese barberry far and wide. In dense stands, Japanese barberry effectively become a monoculture, replacing the diverse, native understory with this invasive species. With its spines and dense growth pattern, it provides a formidable natural barrier or fence. This natural barrier also provides ample habitat to ticks and mice. It is known to be inedible to white-tailed deer and other large mammals, due to the plants lack valuable nutrients. As such there are no mammals in North America that browse and or consume the shrub. Like all barberry species it creates an ample habitat for ticks, leading to and increase in tick population in an area that is infested with Japanese barberry. This can lead to an increase in the likelihood of contracting lime disease.

Economic Impact

As of July of 2020, the Japanese barberry is considered a noxious weed in the state of West Virgina and cannot be bought or sold. In the past Japanese barberry was sold commercially as an ornamental plant famed for its beauty. The main economic impact of Japanese barberry is the cost to remove them from the locations they have infected. Because this species reproduces both sexually and asexually it is of the utmost importance that the entirety of the plant is killed, or uprooted. If one viable root is left underground that could potentially grow into a new barberry bush, which could then spread elsewhere. Also, the treatments and herbicides used to cull and eradicate the Japanese barberry can become expensive very quickly. Due to this species creating an ideal microhabitat for ticks, human health concerns are likely to arise in an area infested with Japanese barberry. As the carriers of the bacteria that cause Lyme disease, increased tick populations could lead to more cases of Lyme and other tick-borne diseases in humans, pets, and livestock. This will cause people bitten by these ticks to spend much money on medical treatments.

Controls and Eradication Measures

- Biological Control

  There are currently no Biological Controls Available to treat this species. (Michigan DNR)

- Mechanical Controls

  There are a few mechanical controls available for this species. The first one being the stump cut method would be to cut it or mow it. You can control the growth of species by mowing it, but the contrast to this method is that it will grow back. This plant cannot be killed with mowing; it can only be cut back and kept it check. If mowing is the control method, it is a constant and will have to be done every so often.

  The next mechanical control method is the digging or pulling method. If this method is chosen, take caution when handling due to the spines on the stem. Handling it with a tool is best so touching the plant with hands is avoided. If handling is required or cannot be avoided proper PPE should be used. Digging up this plant can be done with machinery like an excavator or a backhoe. Although it is more laborious, hand tools like a shovel or mattock can be just as effective. When digging the roots must be pulled up with the plant to kill it and stop the spread. Not removing the roots will allow it to sprout back.

The last mechanical control is a prescribed burn. This control method proves to be effective in removing the plant and slowing the spread. It can be done in combination with an herbicide treatment and another combination of mowing. In combination mowing should be done first and in the herbicide combination the herbicide should be applied after. The burning will mostly kill seedlings and with consistent burns will eventually kill adult or mature species. The most effective time for burning these would be in late spring with a good fuel load in the right conditions. (Michigan DNR) Burning should only be done in a controlled and legal environment. Before burning make sure all your actions are legal and safe. Proper burning usually requires a permit and all burning laws in the state you're in must be followed. All this should be found within the proper agencies, and they should be contacted and/or notified before any action is taken. Burning is the least recommended and should be sought as a last option.

- Chemical Controls

There are two methods for chemical control and that is foliar spray and cut-stump method. Both methods should be taken with precaution and are best done after heavy sap flow in actively growing plants. Whenever using pesticides all precautions should be taken, and all PPE should be worn. All directions and safety instructions should be followed.

A foliar spray is done with a boom applicator or a backpack sprayer. It is applied onto the leaves of the plant. It can be most effective during a drought or period of stress. When applying a foliar spray wind drift and direction should always be paid attention to, so any spray or mist does not come in contact with the applicator or surrounding personnel.

The cut-stump method is done after the plant is cut at the stump and then herbicide is applied directly on to the stump. This is done so that as sap flow is going down it will take the herbicide with it and will kill the plant at the roots. (Michigan DNR)

- Policy Control

As of July 1, 2020, the State of West Virginia has made it illegal to sell Japanese Barberry (*Berberis thunbergii*). (WVDA)

Concept Map

The concept map describes the relationship that Japanese Barbery has with its environment and how it affects the ecosystem and other species. It has been known to house ticks that feed on mammals and birds and the berries that are on this plant feed the birds and mammals. So, when an animal eats the berries, it can give a tick a chance to attach itself to the mammal and/or bird. It will make the soil it grows in more alkaline so that competing native plants will die out leaving more room for the Japanese Barberry to grow, expand, and flourish. Human interaction can have positive and negative effects. The negative effect is dispersal or spread and the positive is treating for its removal and eradication.

Fig. 1 Japanese barberry with leaves (forestryimages.org)

Fig. 2 Close-up shot of Japanese barberry

Fig. 3 Mature Japanese barberry, with fruits

Fig. 4 Area infested with Japanese barberry

References

Kulhanek, Ashley & Kulhanek, A. (n.d.). *Plant introductions: Meet invasive species common barberry and Japanese barberry.* BYGL. https://bygl.osu.edu/node/1726 JapaneseBarberryBCP.pdf

Kurtz, C. (2025, September 9). *Japanese barberry*. Penn State Extension. https://extension.psu.edu/japanese-barberry Japanese Barberry

Kurtz, C. (n.d.). *Northern Research Station*. US Forest Service Research and Development. https://research.fs.usda.gov/nrs An Assessment of Japanese Barberry in Northern U.S. Forests

Mid-Atlantic Writers. (n.d.). *SHRUBS & SUBSHRUBS*. Japanese barberry (Berberis thunbergii). https://www.invasive.org/alien/pubs/midatlantic/beth.htm

Wvda. (2020, June 29). *Sale of Japanese barberry outlawed July 1.* West Virginia Department of Agriculture. https://agriculture.wv.gov/sale-of-japanese-barberry-outlawed-july-1/ Sale of Japanese Barberry Outlawed July 1 : West Virginia Department of Agriculture

Spread of Japanese Honeysuckle (*Lonicera japonica*)

Benjamin Gabehart & Veronica Foster-Toledano

Introduction

Japanese Honeysuckle better known as *Lonicera japonica*, part of the Caprifoliaceae family is native to eastern Asia primarily Japan, Korea and parts of China. It was first discovered and derived from Japan. Japanese Honeysuckle was intentionally introduced into North America and Europe as an ornamental plant and used for erosion control. According to the USDA the introduction into the United States began in the early 19th century because of its fragrant and attractive flowers, rapid growth and utilization as ground cover for stabilization and landscaping. After it was intentionally planted its rapid growth helped it escape cultivation via vegetative spread and seed dispersal by birds and mammals enabling the invasion of wildland. After Japanese Honeysuckle was introduced and established dispersal methods enhanced including vegetative spread. The species formed extensive underground rhizomes and layered stems as well as fragments and lateral rooting which allow them to expand in patches and establish local dominance which is what makes them highly invasive. As noted in the CABI Compendium, *Lonicera japonica* also spread through seed dispersal where birds consume the black berries on the plant and disperse seeds over longer distances, the seeds tend to then establish in open-canopy, forest edges or disturbed habitat. Japanese Honeysuckle was also spread through human mediation such as plantings in gardens, windbreaks and erosion control plantings facilitated spread along roadsides and riparian corridors.

Ecology Impacts

Japanese Honeysuckle is considered an aggressive, invasive vine. It invades forest edges, open woodlands, riparian zones, roadsides, early successional fields and some wetlands. It creates dense mats on the ground and climbs shrubs and smaller trees and smothers them altering the native plant community, structure and function.

The ecological impacts of *Lonicera japonica* are evident in light interception and competition between native species. The vining and climbing growth shades native understory vegetation and suppresses regeneration of tree seedlings and herbaceous by reducing light availability. On the forest floor a thick mat of stems prevents germination and establishment of native plants. According to the CABI Compendium, several field studies documented reduced native species richness and suppressed regeneration with *L. japonica* was present.

Japanese Honeysuckle influences the alteration of successional trajectories. By limiting tree seedling establishment and dominating early successional spaces, Japanese honeysuckle can influence successional pathways, prolonging shrub dominated states or favoring other non-native species. *L. japonica* also affects below ground and soil, dense mats modify litter accumulation, moisture and temperature at the soil surface. Research done and reported in the CABI Compendium shows altered decomposition rates under the invasive vine. The fruit growing on Japanese Honeysuckle is eaten by birds and mammals which aids in seed dispersal. The fruits, however, are low in nutritional value compared to native fruits and berries providing lower quality winter forage for the native animals. Dense mats or tangles in the vine may provide cover for specific species tolerant of the altered plant life however, the overall alteration of plant composition reduces habitat quality for many native invertebrates and plants.

Economic Impact

The control and repeated management of Japanese Honeysuckle through mechanical removal, herbicide treatments and follow up monitoring create direct costs for land managers or private landowners. These reoccurring costs can be substantial on heavily infested public lands, rights-of-way, and restoration sites. Forest landowners also have to swallow the loss of regeneration potential and incur the need for restoration planting. Reduced native plant diversity and forest regeneration can translate into long term losses in ecosystem services such as timber regeneration or native biodiversity which are both difficult to monetize but important both ecologically and economically. Japanese Honeysuckle is most problematic across the eastern United States in colder northern localities its growth can be limited or seasonal while in warmer areas it may remain semi-evergreen and more aggressive. The predicted future model on the US Forest Service website indicates a potential for continued range expansion under specific scenarios and conditions.

Control & Eradication Methods

The eradication of large, long-established infestations of invasive species is not realistic. Management methods usually focus on containment, reduction and restoration. Successful control requires integrated and repeated efforts (mechanical removal, herbicide, monitor, and restore).

Methods of mechanical control include hand pulling and digging, cut-stump treatment, repeated mowing or cutting and mechanical removal of mats. Hand pulling and digging is effective for small infestations and seedlings. You must remove the root crowns and rhizome fragments to prevent resprouting. This method is best to do when the soil is moist to remove the entire root system and is most effective on restoration plots and small sites. Cut-stump treatment works best for larger stems and vines climbing on trees. Cut the stems near the base and immediately apply systemic herbicide (typically a glyphosate or triclopyr formulation labelled for woody stump treatment) to the cut surface to prevent resprouting. This method reduces herbicide use and targets the plant. Repeatedly cutting or mowing above ground vines will deplete the Japanese Honeysuckle carbohydrate reserves overtime if the regrowth is regularly removed (Forestry and Natural Resources, University of Kentucky). The method may take multiple seasons and is labor-intensive and not ideal where vines climb trees as it risks girdling. Mechanical removal of mats and rhizomes is effective on accessible sites. Excavating roots can be effective however it disturbs the soil and non-target species. Mechanical removal should also be followed by restoration planting and erosion control. Mechanical methods are labor intensive and effective primarily on a small scale, without follow up treatment's plants resprout and reestablish.

Biological control is not well identified or established for Japanese Honeysuckle in North America. Biological-control research focused on identifying host-specific agents, but the species' close relation to ornamental or beneficial species and the risk of non-target impacts make introduction of biological control agents challenging. A common recommended biological approach is restoration with native shrubs and trees that shade out seedlings and occupy niches this reduces the reinvasion pressure of *L. japonica*. Increasing the canopy closure reduces the honeysuckles vigor because the vine prefers higher light. Establishing dense native groundcovers and competitive plants can be a long-term suppression strategy. Any prospective insect or pathogen control agent requires rigorous host specificity testing. Until a safe agent is tested and identified, management relies on mechanical and chemical tools and restoration.

Chemical control most commonly consists of herbicide, the two main forms being glyphosate and triclopyr. Glyphosate is systemic, non-selective and mostly applied as a foliar spray on actively growing foliage or applied to cut-stumps. The care needed to limit non-target exposure, and application timing on active growth influences efficacy. Triclopyr selective on woody broadleaves is preferred in areas where preservation of grasses is desired. It is used as basal bark, cut stump or foliar treatment depending on the formulation. Application strategies vary depending on the size, age, and location of the plant. Foliar spray is most effective on young to medium size plants in leaf, requires thorough coverage. Avoid spraying on desirable plants, follow all precautions and note if multiple treatments are necessary. Cut stump and basal bark spray is best used for large vines and stumps. After cutting apply herbicide to the stumps surface to translocate the roots and prevent resprouting. Basal bark applications are used on the stems and concentrate herbicide use and reduce non-target exposure. Herbicides can be effective but require careful timing, appropriate formulations, and follow-up. Repeated treatments are usually required because root and rhizome stores and seed banked recruits can produce new shoots. Using herbicides near water requires formulations and methods that meet label restrictions. Always follow local regulations and label instructions.

Practical protocol starts with assessing the site and setting goals. Determine whether eradication, containment or suppression is realistic based on the abundance and size of the Japanese Honeysuckle. Prioritize high value conservation areas with the most intensive control. For a dense infestation a combined mechanical cut and targeted herbicide is typically effective to quickly reduce above ground biomass while limiting spraying non-target species. A follow-up treatment should consist of monitoring the site after the cut for resprouts and seedlings and treat emergent plants with spot herbicide or hand pull small plants. Follow up should last multiple years and continue during restoration and planting period. Replant with fast growing native shrubs, trees and dense ground coverage to close the canopy faster preventing the regrowth of *L. japonica*. Prevent new plantings of Japanese Honeysuckle, discourage the use of ornamentals and promote native alternatives.

Japanese honeysuckle (*Lonicera japonica*) was an intentionally introduced ornamental and erosion-control vine that has become invasive across much of North America. It spreads via both vegetative growth and bird-dispersed seed. It forms dense ground mats and climbing tangles that suppress native plants and alter successional trajectories. It imposes recurring management costs and effective management requires integrated, site-specific approaches (mechanical removal, targeted herbicide use, and ecological restoration) with long-term follow-up and prevention measures.

Photos

Fig. 1 Left: This photo depicts Japanese Honeysuckle during the growing season. It has long slender stems, leaves that are oval and oblong, typically 1-3 inches with smooth to slightly hairy texture and bright to dark green color and small shiny black berries that ripen in clusters Right: This photo shows Japanese Honeysuckle in the dormant season. The leaves are yellowing and will begin to drop. The stems become more visible and are a reddish brown to light tan color. Berries become sparse and will eventually drop too.

Concept Map

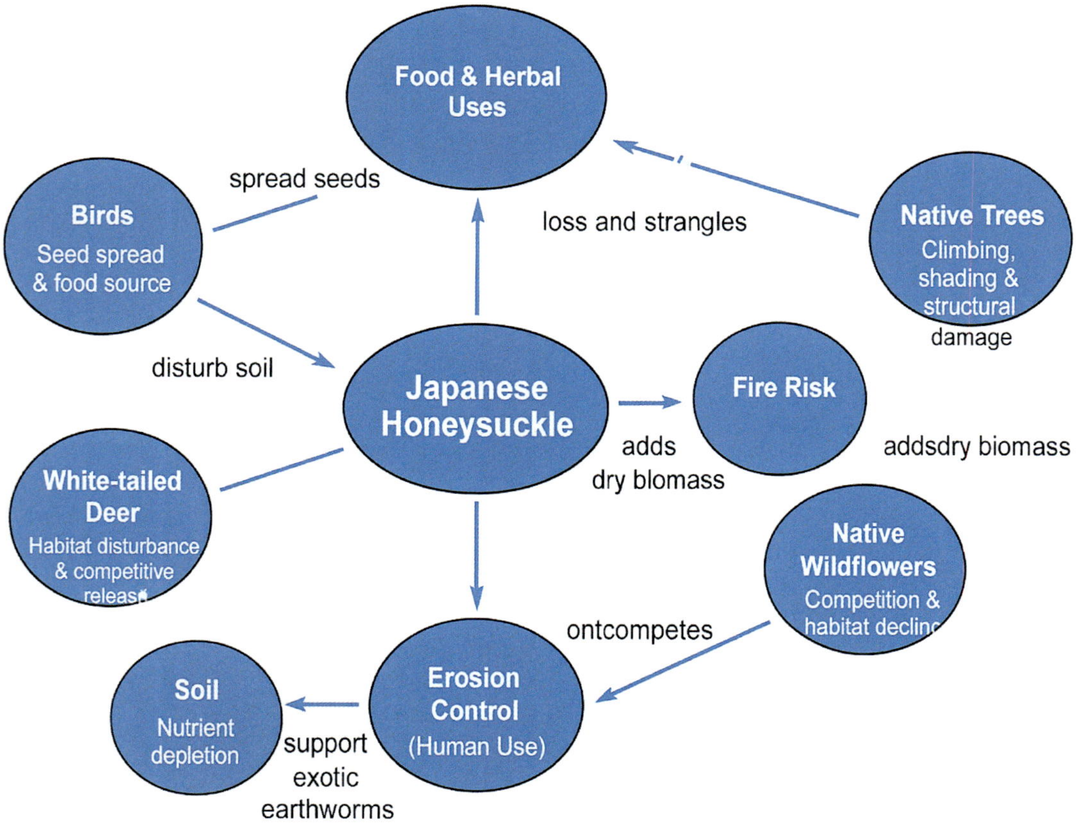

Concept Map Explanation

Birds (Native and Invasive) – Seed Spread and Food Source

Robins and European starlings play a key role in the spread of Japanese honeysuckle by consuming its nutrient-rich berries. After feeding, these birds disperse the vine's seeds through their droppings, allowing honeysuckle to colonize new areas, including undisturbed forests. This process enables the species to invade habitats that would otherwise be difficult to reach. While the berries provide a valuable winter food source, particularly when other resources are scarce, this interaction disrupts native plant reproduction. Birds become less reliant on native berry-producing plants, indirectly reducing native biodiversity and altering natural food web dynamics. Japanese honeysuckle provides nourishment for birds, while birds in turn spread its seeds and facilitate its expansion.

White-Tailed Deer – Habitat Disturbance and Competitive Release

White-tailed deer interact with Japanese honeysuckle in ways that unintentionally encourage its growth. Although deer browse on honeysuckle leaves and shoots, the plant's ability to quickly resprout allows it to recover rapidly. Deer prefer native plants, and their heavy grazing pressure on native wildflowers, shrubs, and tree saplings reduces the native plant community's ability to compete. This over-browsing opens light gaps in the forest understory and disturbs the soil, both of which provide favorable conditions for honeysuckle growth. Additionally, deer hooves physically disturb soil surfaces, helping spread honeysuckle seeds and root fragments. In this mutual but imbalanced relationship, deer promote honeysuckle's expansion through habitat disturbance, while the vine provides browse and cover in return.

Native Trees – Climbing, Shading, and Structural Damage

Japanese honeysuckle poses significant threats to native trees by twining tightly around their trunks and branches. As it climbs, the vine spreads thick layers of foliage that block sunlight from reaching the host tree's leaves, reducing photosynthesis and stunting growth. Over time, the added weight of mature vines can break limbs, increasing vulnerability to wind and storm damage. In younger forests, the vine can entirely smother saplings before they mature, leading to a decline in native canopy formation. As large trees weaken and die, more light penetrates the forest floor, creating ideal conditions for honeysuckle to grow even more aggressively. This creates a self-reinforcing cycle of tree loss and invasive dominance.

Native Wildflowers and Understory Plants – Competition and Habitat Decline

Native wildflowers, ferns, and shrubs are among the species most affected by Japanese honeysuckle invasion. The vine forms dense mats that spread across the ground, shading out low-growing vegetation and monopolizing soil nutrients and moisture. These dense growths prevent sunlight from reaching the understory, causing declines in native plant diversity. As native wildflowers and shrubs disappear, pollinators such as bees and butterflies lose vital nectar sources, and other wildlife that depend on native plants for food or shelter are similarly impacted. Over time, this competition leads to simplified plant communities and widespread biodiversity loss in forest understories.

Soil – Nutrient Depletion and Indirect Effects

Japanese honeysuckle also alters the soil environment where it grows. Its dense root and leaf mats extract large amounts of moisture and nutrients, leaving the soil depleted and less fertile for native species. The vine changes the microclimate near the ground, affecting decomposition rates and soil temperature. These changes can favor exotic earthworms, which further disrupt soil structure and accelerate leaf litter breakdown. The combined effects reduce the amount of organic matter available for native plant seedlings to establish, leading to long-term declines in soil health and native plant regeneration.

Erosion Control (Human Use) – Mixed Impacts

Japanese honeysuckle was originally introduced to North America for use in erosion control and ornamental landscaping. Its dense root system effectively stabilizes soil on steep slopes, roadsides, and other disturbed areas, providing short-term erosion prevention. However, these same characteristics allow the plant to spread uncontrollably into adjacent natural areas. Once established, it often outcompetes native grasses and shrubs that would naturally recolonize disturbed ground, forming monocultures that suppress biodiversity. Therefore, while honeysuckle can help stabilize soil, it does so at the cost of long-term ecological balance and native plant recovery.

Food and Herbal Uses (Human Use)

Humans have found various uses for Japanese honeysuckle, particularly in traditional medicine and herbal practices. The flowers and leaves are used to make teas and extracts, believed to have anti-inflammatory and antimicrobial properties. The nectar is also edible and has been used in beverages and flavorings. Despite these beneficial uses, the intentional cultivation or wild harvesting of honeysuckle contributes to its continued spread, as seeds and plant fragments can easily escape into surrounding habitats. These practices, although culturally or economically valuable, further increase the risk of ecological harm by promoting the plant's proliferation beyond controlled environments.

Fire Risk and Habitat Alteration

As Japanese honeysuckle grows and matures, it produces dense layers of vegetation that dry out during late summer or winter, increasing the accumulation of flammable material in forest understories and along roadsides. These vine tangles act as fuel, raising the risk and intensity of wildfires. After a fire, honeysuckle has a strong advantage over native plants due to its ability to resprout quickly from surviving roots and rhizomes. This rapid regrowth allows it to dominate post-fire landscapes, suppressing native plant recovery and altering habitat structure. Honeysuckle not only contributes to higher fire risk but also benefits from the disturbances it helps create.

References

Bush Honeysuckle - Extension Publications, publications.mgcafe.uky.edu/files/FOR175.pdf. Accessed 15 Nov. 2025.

"EPPC - Southeast Exotic Pest Plant Council." *Tree Strangling & Canopy Takeover*, www.se-eppc.org/. Accessed 14 Nov. 2025.

"Forest Floor Domination & Loss of Natives ." *Japanese Honeysuckle (Lonicera Japonica Thunb.)*, www.invasive.org/browse/subinfo.cfm?sub=3039. Accessed 14 Nov. 2025.

"Japanese Honeysuckle." *Japanese Honeysuckle | National Invasive Species Information Center*, www.invasivespeciesinfo.gov/terrestrial/plants/japanese-honeysuckle?utm.com. Accessed 14 Nov. 2025.

"Japanese Honeysuckle Control." *Missouri Department of Conservation*, mdc.mo.gov/trees-plants/invasive-plants/japanese-honeysuckle-control?utm. Accessed 14 Nov. 2025.

"Join the Movement: NWF Native Plant Habitats." *National Wildlife Federation*, www.nwf.org/Garden-for-Wildlife/About/Invasive-Species. Accessed 14 Nov. 2025.

"Missouri Botanical Garden – Growth Habit and Harm to Trees." *Plant Finder*, www.missouribotanicalgarden.org/PlantFinder/PlantFinderDetails.aspx?taxonid=281091. Accessed 14 Nov. 2025.

Munger, Gregory T. 2002. Lonicera japonica. In: Fire Effects Information System, [Online]. U.S. Department of Agriculture, Forest Service, Rocky Mountain Research Station, Fire Sciences Laboratory (Producer). Available: https://www.fs.usda.gov /database/feis/plants/vine/lonjap/all.html [2025, November 14].

"Plant Guide Plant Materials." *USDA NRCS Plant Guide – Deer Browsing Encourages Spread*, plants.sc.egov.usda.gov/DocumentLibrary/plantguide/pdf/cs_losi.pdf. Accessed 14 Nov. 2025.

"Plant Invaders of Mid-Atlantic Natural Areas." *National Park Service – Disturbance Favors Japanese Honeysuckle*, www.invasive.org/eastern/midatlantic/loni.html. Accessed 14 Nov. 2025.

Rojas-Sandoval, Julissa and Acevedo-Rodríguez, Pedro, cabicompendium.31191, CABI Compendium, doi:10.1079/cabicompendium.31191, CABI, Lonicera japonica (Japanese honeysuckle), (2022)

Stearns, Stacey. "Invasive Plant Factsheet: Japanese Honeysuckle." *Integrated Pest Management*, 23 Aug. 2021, ipm.cahnr.uconn.edu/invasiveplantsjapanesehoneysuckle/.

"Wildlife Interactions and Spread." *US Forest Service*, 29 Sept. 2025, www.fs.usda.gov/.

Understory Control and Management of Japanese Stilt Grass (*Mircostegium vimineum*)

Joshua Weister & Sam Turley

Species Name

The species *Microstegium vimineum* was originally named *Andropogon vimineus* by a German physician and botanist named Karl Bernhard von Trinius in 1832 (USFWS 2018). Christian Gottfried Daniel Nees von Esenback later named the species *Microstegium willdenowianum*. This, in 1836, was when the species was classified into the *Microstegium* genera. The genera, *Microstegium,* is of Greek origin and means "small roof." *Vimineum* is of Latin origin and means "of wickerwork." The current name, *Microstegium vimineum* was given to the species in 1922 by French botanist Aimée Antoinette Camus. Camus is well known for being the second most prolific female taxonomist. In total, she delineated six hundred and seventy-seven species, including *Microstegium vimineum.* (USFWS 2018)

Native Range

Microstegium vimineum is native to Far-East Asia, specifically Japan, Korea, China, Malaysia, India, and the Caucasus Mountains. The climate of this area ranges between temperate and warm continental climates (Fryer 2011). This is a very similar climate to that found in the eastern US, which is likely a factor to why Japanese Stiltgrass does so well in the Eastern US. The species has also reached and invaded other parts of Asia, including Pakistan, Nepal, and Turkey (Fryer 2011). In the US, stiltgrass is commonly found in areas such as forest edges, wetlands, and disturbed sites. Domestically in the US, stiltgrass ranges from Texas, its westernmost boundary, to New York, its easternmost boundary. It is fairly reasonable to assume stiltgrass can be found in these areas in its native range as well. Anecdotally, I once noted seeing Stiltgrass in the Philippines, where it was growing as what looked like a weed in the landscaping of a restaurant parking lot.

Method of Introduction

According to Fryer (2011), stiltgrass was likely introduced accidentally to the Eastern US as a packing material, specifically packing material for fine Chinese porcelain. Packing material would typically consist of densely packed stalks of cheap or otherwise useless vegetation that was widely abundant, which perfectly describes Japanese Stiltgrass. The packing material would simply be discarded after unpacking, leaving the material laying out wherever the items were unpacked, and therefore creating a colony in that area. This was a common practice in the US in the early 20th century, as several different invasive species have been introduced in this way. Stiltgrass spread rapidly in this way due to its fast seed development and high reproductive capacity.

Characteristics

Japanese stiltgrass has many ecological impacts. When Japanese stiltgrass grows, it increases the soil pH and phosphorus which can reduce the potential diversity of the soil. Japanese stilt grass can accelerate the carbon cycling and deplete the soils below-ground carbon pools in forests. Due to its dense pad growth, it can reduce native tree seedling regeneration by blocking the seeds of native trees and other vegetation from reaching the forest floor to regenerate. This can reduce the rate of succession in forests and alter the species composition of the forests.

Japanese stiltgrass, like many other invasive species, impacts economies by harming the regeneration of commercially-viable tree seedlings. Japanese stiltgrass is especially hard for competing species to overcome, because it forms such a dense cover of the ground that it can be very hard for any new seedlings to receive enough light to get properly established. Without proper regeneration, the site will eventually lose timber value as the existing canopy level trees age out or are harvested. The stiltgrass will prevent new growth from occurring, meaning that once the timber harvested or becomes too old and poor quality to harvest, there is no younger growth to replace it. This can significantly reduce income for landowners.

Another economic impact of Japanese Stiltgrass is its effect on the fire regime. Because it is an annual grass, Japanese Stiltgrass leaves behind a large amount dead biomass each year. This can lead to an extra buildup of fuel load, leading to more wildfire risk. This can impact landowners in several ways. First, if a fire were to start, the stiltgrass could contribute to it becoming even more powerful, which would be more damaging to the landowner's financial state. Increased fuel loads may require landowners to spend additional money on fuel reduction and management to protect their property and prevent highly damaging fires. This can further reduce landowner income.

Control and Eradication Measures

Japanese stilt grass can be spread very easily and become a nuisance very easily. Some ways to control stilt grass are mechanical and chemical. Stilt grass normally spreads to disturbed areas of land. Stilt grass is an annual plant, it creates a seed bank, so it will grow back in the same area each year if left untreated. Mechanical treatment can consist of regular mowing to reduce the seed production. Mowing in the late summer can be effective since the seeds are ready to be produced then. Hand weeding is effective when it comes to small patches, but not practical for large patches. Some situations where the area is moist, flame weeding is effective if the fire hazard can be controlled. The best time for mowing to reduce seed production is before the flowers are visible. Once other plants have started flowering, the stilt grass seeds are mature and will persist even if the grass is mowed.

Fig. 1 This is a picture of Japanese stilt grass taken outside of the apartment we live in. It has spread all over the disturbed areas under the tree canopy in this area.

Chemical control is another effective treatment for Japanese stilt grass. Herbicides that are labeled as preemergence and postemergence will control Japanese stilt grass. In landscape settings, indaziflam and dithiopyr are labeled as preemergence and will control stiltgrass. To control Japanese stiltgrass for selective postemergence, clethodim, sethoxydim, and fenoxaprop are effective. Also, low doses of glyphosate are effective. When using these chemicals for herbicide applications, applications must be made before the seeds germinate so the Japanese stiltgrass doesn't germinate and deposit their seeds into the seed bank for the next year. There are no current biological control treatment methods for Japanese stiltgrass.

Fig. 2 This is a picture of Japanese stilt grass taken outside of the apartment we live in. It has spread all over the disturbed areas under the tree canopy in this area.

Fig. 3 This is Sam examining the Japanese Stiltgrass.

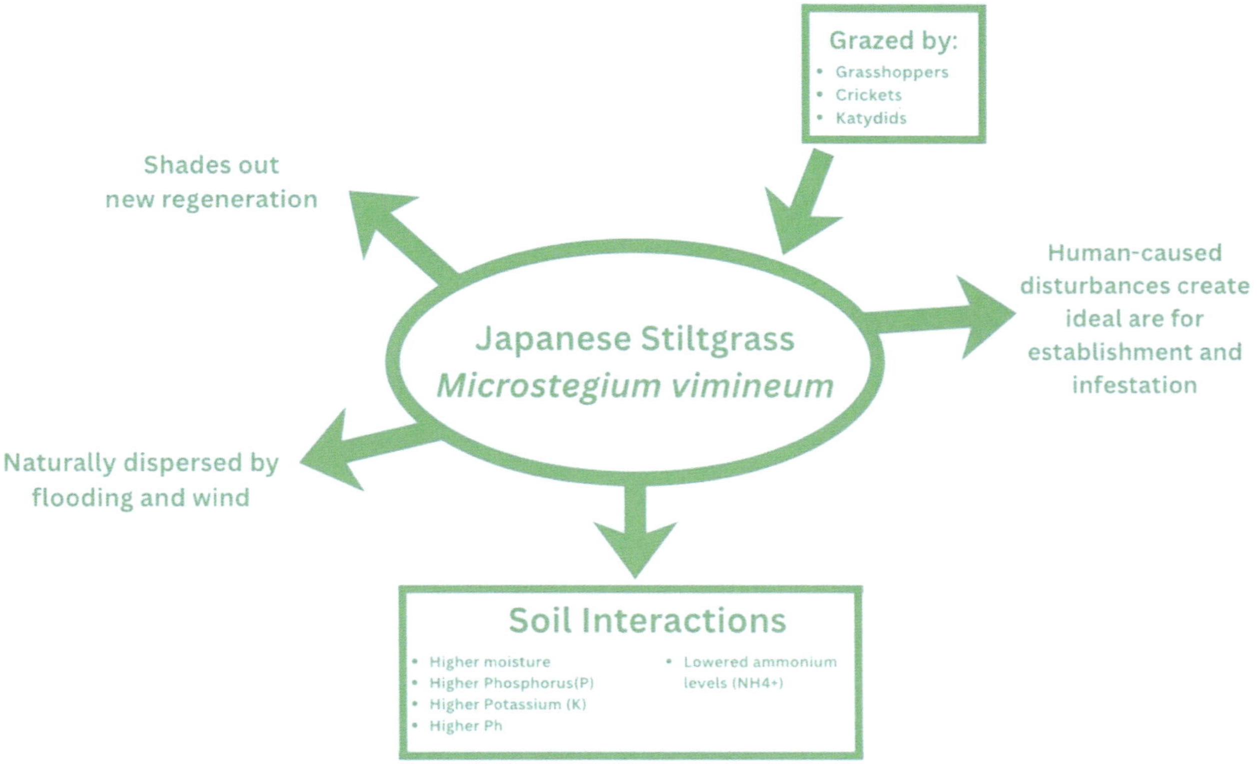

The Concept Map above is a flow chart showing how Japanese stiltgrass interacts with its environment. As it grows, it forms a thick, dense mat that can trap and retain nutrients. Stiltgrass seeds are easily dispersed by wind and water, which helps it spread. Flooding indirectly spreads the siltgrass seeds. Humans and machines also indirectly spread the seeds by getting stuck on the boots of workers and the wheels and parts of the machines. White-tailed deer also indirectly spread Stiltgrass. Deep leaf litter inhibits the spread of stiltgrass by smothering out its ability to grow. Short-horned grasshoppers, crickets, and katydids all graze on stiltgrass.

References

1 Japanese stiltgrass (Microstegium Vimineum) Available at: https://www.fws.gov/sites/default/files/documents/Ecological-Risk-Screening-Summary-Japanese-Stiltgrass.pdf. (Accessed: 17th November 2025)

Culpepper, L. Z., Wang, H.-H., Koralewski, T. E., Grant, W. E. & Rogers, W. E. Understory upheaval: Factors influencing Japanese stiltgrass invasion in Forestlands of Tennessee, United States. Botanical studies (2018). Available at: https://pmc.ncbi.nlm.nih.gov/articles/PMC6079111/. (Accessed: 16th November 2025)

Defining autopolyploidy: Cytology, genetics, and taxonomy Available at: https://bsapubs.onlinelibrary.wiley.com/doi/epdf/10.1002/ajb2.16292. (Accessed: 17th November 2025)

Effects of a non-native grass invasion decline over time - flory - 2017 - journal of ecology - wiley online library Available at: https://besjournals.onlinelibrary.wiley.com/doi/pdf/10.1111/1365-2745.12850. (Accessed: 17th November 2025)

Egrove. Site Available at: https://egrove.olemiss.edu/. (Accessed: 16th November 2025)

Elevational variability in and controls on the temperature sensitivity of soil organic matter decomposition in alpine forests Available at: https://esajournals.onlinelibrary.wiley.com/doi/epdf/10.1002/ecs2.4010. (Accessed: 17th November 2025)

Habitat, dispersal and propagule pressure control exotic plant infilling within an invaded range - warren - 2013 - ecosphere - wiley online library Available at: https://esajournals.onlinelibrary.wiley.com/doi/pdf/10.1890/ES12-00393.1. (Accessed: 17th November 2025)

Japanese stiltgrass. Maine Natural Areas Program, Invasive Plants, Japanese stiltgrass Available at: https://www.maine.gov/dacf/mnap/features/invasive_plants/microstegium.html. (Accessed: 16th November 2025)

Landsman, A. P., Burghardt, K. T. & Bowman, J. L. Invasive grass (microstegium vimineum) indirectly benefits spider community by subsidizing available prey. Ecology and evolution (2020). Available at: https://pmc.ncbi.nlm.nih.gov/articles/PMC7593186/?utm_source=chatgpt.com. (Accessed: 16th November 2025)

Lucid visual collaboration suite: Log in Available at: https://lucid.app/lucidchart/0ead1c3c-d8cb-4d28-8ff0-63b24c14e1fb/edit?beaconFlowId=260B0090ECBF85A8&invitationId=inv_693b789b-4bbe-4bbc-9629-d149730c4e57&page=0_0#. (Accessed: 16th November 2025)

Microstegium Vimineum in Flora of china @ efloras.org Available at: http://www.efloras.org/florataxon.aspx?flora_id=2&taxon_id=200025707. (Accessed: 16th November 2025)

Neal, J. & Judge, C. A. Japanese stiltgrass identification and management: NC State Extension Publications. Japanese Stiltgrass Identification and Management | NC State Extension Publications Available at: https://content.ces.ncsu.edu/japanese-stiltgrass-identification-and-management. (Accessed: 16th November 2025)

Nestory, S. Biological and mechanical control of Japanese stiltgrass (Microstegium Vimineum). DSpace Repository (1970). Available at: https://udspace.udel.edu/items/6e481d93-e21f-4d15-a7f8-66a2759eb6d3?utm_source=chatgpt.com. (Accessed: 16th November 2025)

Payne, G., Evans, J., Murdock, E. & Derr, J. Comparison of torching and glyphosate applications for Japanese stiltgrass (Microstegium Vimineum) control. Journal of Environmental Horticulture (2022). Available at: https://jeh.kglmeridian.com/view/journals/jenh/40/4/article-p149.xml?utm_source=chatgpt.com. (Accessed: 16th November 2025)

PMC Home. National Center for Biotechnology Information Available at: https://pmc.ncbi.nlm.nih.gov/. (Accessed: 16th November 2025)

www.upane.it, U.-. GISD Available at: https://www.iucngisd.org/gisd/speciesname/Microstegium+vimineum. (Accessed: 16th November 2025)

Overview of Kudzu's (*Pueraria lobate*) Morphology, Economic and Ecologic Impact

Spencer Nolan

Introduction

Pueraria lobata commonly known as Kudzu, is currently an invasive species within much of the southern United States and is currently a growing problem within West Virginia. Kudzu is a particularly damaging invasive species, due to its high rate of growth and being a vine, allowing it to smother native species and restrict access to sunlight. Kudzu is one of the most famous invasive species within the Southeastern United States, to the point it is now ingrained within the culture of the region.

Morphology

Kudzu is a climbing, semi-woody vine, which is alternately arranged with pinnately compound leaf that contains three leaflets. Flowers are a typical pea construction and are located on a raceme. These purple flowers bloom in late summer. The fruits are a pea-like pod with brown pubescence and are typically around 2 inches long. Kudzu has roots that can grow 1-3 m (3-9 ft) and are tuberous, up to 17.8 cm (7 in) in diameter (Bargeron, n.d.). Kudzu typically prefers disturbed sites such as roadsides and abandoned fields. Kudzu is most notably known for its fast rate of growth, and it is not uncommon for it to grow upwards of 50 ft in a single year. Kudzu is also a legume, meaning it fixes nitrogen into the soil. Kudzu most commonly spreads and reproduces clonally through its rhizomes, allowing it to take over sites quickly. Kudzu also produces the compounds genistein, daidzein, and puerarin, which are believed to be allelopathic (Rashid, Asaeda, Uddin, 2010).

Fig. 1 Kudzu leaf in late fall, with light defoliation

Fig. 2 Kudzu "beans"

Species History

Kudzu is native throughout China and Japan but, was first introduced into the United States in 1876 at the United States Centennial Exposition in Philadelphia, Pennsylvania. (Blaustein, 2001) By the 1930's Kudzu became a common plant for erosion control and even was used for grazing livestock. Kudzu was also used for hay and as a cover crop. It was believed by many at the time that Kudzu would never become invasive stating that "Such ideas are unfounded. It may be easily eradicated by grazing or by plowing" (Sturkie, Grimes, 1939) However, as Kudzu began to spread as an invasive, it became less attractive to plant, and by 1970 the USDA lowered the plants status to a common weed, and in 1998, was listed as a noxious weed (Sikora, 2014). Kudzu is now a very prominent invasive within the southern United States, to the point it is ingrained in culture, being known as the "Vine that ate the South". While Kudzu may not be as culturally significant in other areas of the United States, its range still extends north to New York and even has a presence in Oregon and Washington.

Economic and Ecological Impact

Kudzu negatively impacts both the economy and ecology of the United States. Kudzu can carry the pathogen that causes soybean rust over the winter. Soybean rust can lead to reduced yields by reducing the number of pods produced and the number of seeds within pods. While kudzu does not currently have a major foothold within the Midwest, it is estimated that soybean losses could exceed 80% (Sikora, 2014). Soybeans are one of the leading crops in terms of economic output for the American economy, and a yield reduced by that margin would be detrimental to agriculture within the United States. Kudzu is also projected to negatively impact the timber industry, in Oklahoma, its projected that invasion of kudzu into saplings could cost the Oklahoma timber industry $43.36 million in 5 years (Harron, Joshi, Edgar, Paudel, Adhikari, 2020). Kudzu being able to shroud out future timberland is not just a economic issue, but also an ecological one, as a monoculture mat of kudzu can lead to reduced native biodiversity and possibly reduced carbon sequestration over long periods of time. Kudzu has also been linked to an increase in nitric oxide emissions, in one extreme simulation, kudzu was linked to a 28% spike in nitric oxide emissions (Hickman, Wu, Mickley, Lerdau, 2010). Nitric oxide emissions are a greenhouse gas that is a contributing factor to smog and acid rain. Kudzu also causes around $1.5 million in damages to power companies, who have to constantly remove kudzu from power lines (Blaustein, 2001).

Controls

Kudzu can be controlled in a variety of ways, one of the most accessible for landowners and farmers being livestock browse, because kudzu is a legume, it is high in nitrogen content, so it is important to have the proper amount of livestock per acre, to ensure adequate competition to prevent nitrogen poisoning. Browsing by livestock can be a fairly effective strategy when on abandoned pasture or otherwise accessible land, but becomes impractical when on undesirable terrain, or when kudzu has already successfully invaded the canopy. Kudzu can also be controlled with the use of herbicide, when carrying out a cut-stump treatment, it is best to use a 25% Glyphosate solution or a 25% Triclopyr solution. Foliar sprays are best with a 2% Glyphosate and 0.5% surfactant solution or a 2% Triclopyr and 0.5% surfactant solution. Cut-stump treatments are best for killing vines that have spread into the canopy, while foliar sprays are best to control large areas of kudzu (Bargeron).

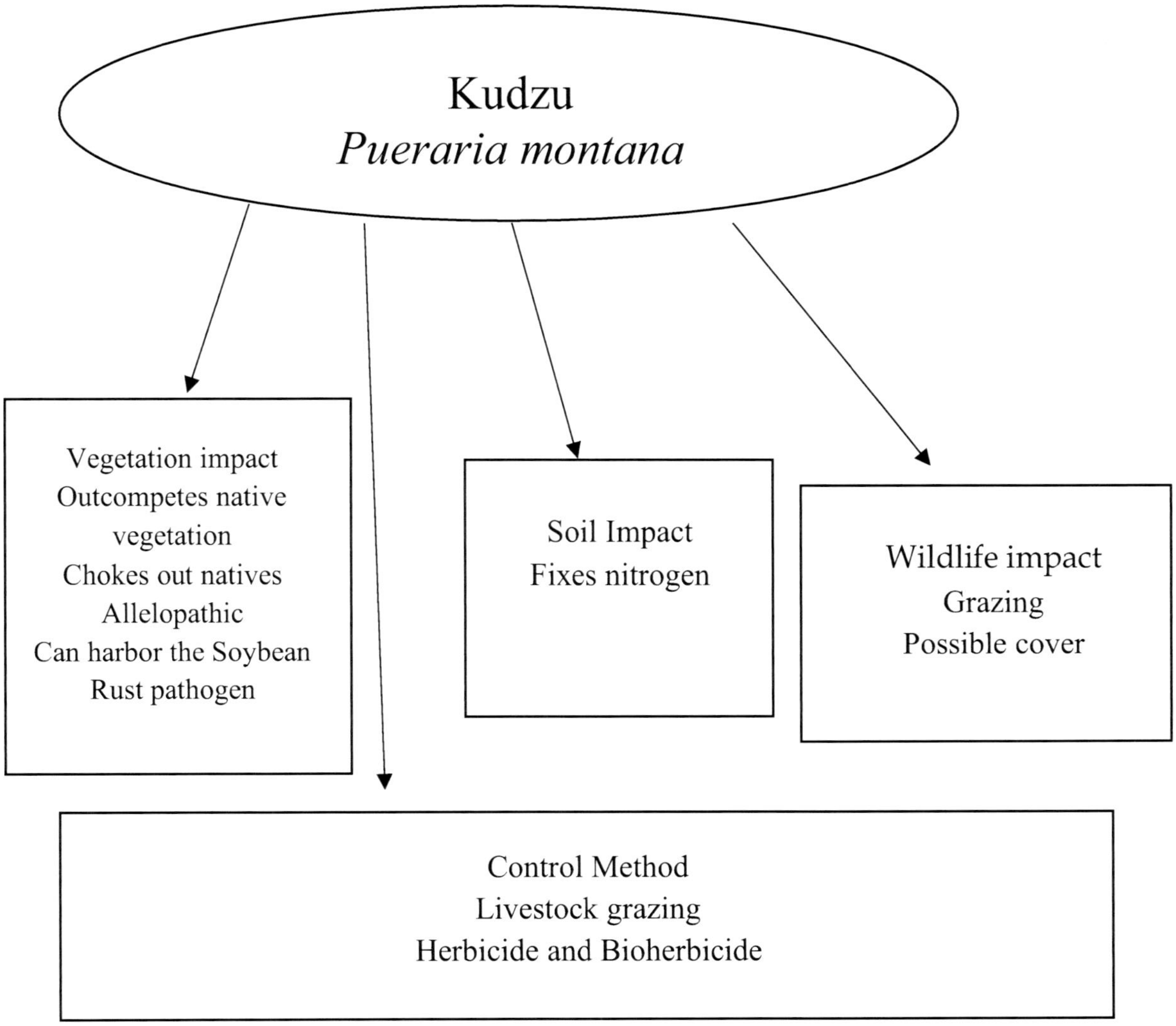

Above is a concept map for kudzu. This shows how kudzu impacts a variety of ecological factors. Kudzu affects vegetation by outcompeting and choking out native species, possibly being allelopathic, and harboring the soybean rust pathogen. Kudzu influences wildlife by providing browse/grazing opportunities and cover. Kudzu can be controlled through livestock grazing and the use of herbicides.

References

Bargeron, C. T. (n.d.). *Southeast Exotic Pest Plant Council Invasive Plant Manual.* Kudzu, pueraria Montana - Southeast Exotic Pest Plant Council Invasive Plant Manual. https://www.se-eppc.org/manual/kudzu.html

Blaustein, R. J. (2001). Kudzu's invasion into Southern United States life and culture. https://www.srs.fs.usda.gov/pubs/ja/ja_blaustein001.pdf

Hickman, Jonathan E., et al. "Kudzu (*Pueraria montana*) invasion doubles emissions of nitric oxide and increases ozone pollution." *Proceedings of the National Academy of Sciences* 107.22 (2010): 10115-10119.

Harron, Paulina, et al. "Predicting Kudzu (*Pueraria montana*) spread and its economic impacts in timber industry: A case study from Oklahoma." *PloS one* 15.3 (2020): e0229835.

Rashid, Md H., Takashi Asaeda, and Md N. Uddin. "The allelopathic potential of kudzu (*Pueraria montana*)." *Weed Science* 58.1 (2010): 47-55.

Sikora, Edward. "Kudzu: Invasive weed supports the soybean rust pathogen through winter months in Southeastern United States." *Outlooks on Pest Management* 25.2 (2014): 175-179.

Forest Damage and Management of Multiflora Rose (*Rosa multiflora*)

Zachary Holstion & Jackson Mallett

Species History

Multiflora Rose (*Rosa multiflora*) is a shrub that is native to Eastern Asia. It was introduced to America in the mid to late 1800s for several different reasons. First of all, it was introduced to America as a rootstock for ornamental roses. It was brought in for rootstock because of its white to pinkish colored flowers, to be used to graft other roses onto to create different types of roses. Secondly, in the 1930s, the US Soil Conservation Service promoted the use of Multiflora Rose as a living fence for multiple different reasons. It can be used to keep livestock in pasture fields so that a traditional manmade fence is not needed. This would mean that farmers would no longer need to build a wire or wood fence, dig holes, or set posts to keep livestock in a desired area. Next, Multiflora Rose was introduced to be used as a method of erosion control or prevention. Its dense thickets can produce very dense root systems that help keep soil stable and in place. Another reason for the introduction of Multiflora Rose to America was the hope that it would be beneficial to wildlife habitats. It was previously thought to be beneficial to wildlife because its dense thickets could provide cover and shelter for birds and small mammals, and its rosehips could provide a food source to various animals.

Characteristics

Multiflora rose (*Rosa multiflora*) is a perennial shrub that is part of the Rosaceae (Rose) family. They are a deciduous shrub that can grow anywhere from 2-10 feet tall. They grow in very dense thickets on roadsides, forest edges, open fields, and disturbed sites. The stem of the plant varies in color depending on the time of the year and the age. The stem can be green to a light reddish color, and over time, with age and being a woody shrub, they can turn into a woody brown color. The stem of the plant is covered with thorns that protect the plant from being eaten by herbivores. The thorns are a green to brownish color, stout in size, and curved, giving them a shape like a shark's tooth. The leaves of the Multiflora Rose are compounded, with 5-11 leaflets. The leaves are oval shaped with many fine teeth. The leaves have an average length of about an inch. Multiflora Rose has a white to pinkish flower that blooms from May to June. The flowers have five petals that have a notch on the edge of the petals, giving them a heart shape. The flowers bloom in a cluster that makes them seem like one big flower. Multiflora rose produces a fruit called rose hips. Rose hips are a small berry-like shape that is approximately a quarter of an inch in diameter, and they are a bright red color. These fruits are eaten by birds, which are known to spread the seeds all over.

Ecological & Economic Impacts

Multiflora Rose poses many ecological impacts in the United States. First off, its rapid growth affects several different parts of the ecosystem. It can outcompete native flora by choking out their roots and forming dense thickets that can pose a challenge for native species to penetrate. These thick root systems can take soil nutrients and moisture from native plants. By choking out native flowering plants, Multiflora Rose can reduce floral diversity and disrupt the foraging patterns of pollinators. The dense thickets of Multiflora Rose can also impede natural wildlife movement, causing animals such as deer to travel routes that they wouldn't take in an ecosystem dominated by native plants. Multiflora Rose can also greatly alter forest succession by dominating the understory. It can suppress the regeneration of native tree seedlings, which alters natural forest development. Multiflora Rose has also been known to hybridize with native rose species, potentially threatening their genetic integrity. Another ecological impact would be the alteration of fire dynamics that the Multiflora Rose presents. Due to its high biomass and very dense structure, it can be a hazard in the event of a forest fire.

Habitat

Due to the aggressive invasive nature of Multiflora Rose, it is classified as a noxious weed in several US states, including Iowa, Ohio, West Virginia, and New Jersey. Multiflora Rose can be very difficult to control once it is introduced. Although there are several ways to control or prevent it, none of them have been proven to be 100% successful. This plant is a shade-tolerant species that can grow under canopies. Even though they are a shade-tolerant species, they prefer to have full sunlight, so they can have the maximum amount of energy to grow and reproduce the plant more in other areas. This plant can be found in a variety of different types of locations, including roadsides, pastures, open fields, forest edges, and disturbed sites such as timber cuts or food plots. Multiflora Rose can grow in a variety of soils ranging from wetlands to drylands and from clay soil to sandy soil. They are found commonly in West Virginia, no matter the soil type or the topography. One of the main causes of Multiflora Rose residing in varying habitat locations is wildlife spreading the seed to a different area, altering the habitat that is already there. Birds and deer will browse the plants if necessary and will disperse the seed from the host of the plant.

Mechanical Control

In terms of mechanical control, Multiflora Rose can be repeatedly mowed or cut down during the growing season as an effort to suppress regrowth and exhaust root reserves. It can also be pulled by hand or dug up if the soil is moist and the infestation is still small. The entire root system must be removed for this method to be effective. Cutting or pruning is another mechanical way to control Multiflora Rose. By cutting the stems close to the ground, you can reduce vigor, but this method typically requires follow-up treatments to prevent regrowth. In the event of heavy infestation, thickets can be bulldozed, although this could potentially cause soil disturbance and/or erosion. Flaming or heat treatment can be used to kill young plants or regrowth, although it must be done very cautiously to avoid forest fires. When suppressing the invasive species with flames, you will need a hot, intense fire that will kill the plant and its seeds in the seed bank, thereby disrupting future regeneration of the plant.

Biological Control

When it comes to biological control options, Rose Rosette Disease (RRD) is the most promising control agent. Rose Rosette Disease is a naturally occurring viral disease spread by the eriophyid mite. It is known to cause abnormal growth and the eventual death of Multiflora Rose plants. Goat grazing is another biological control option. Goats can be very effective at browsing and suppressing Multiflora Rose growth. While goat grazing is not technically a true biological control, their selective feeding behaviors can mimic natural herbivory and can reduce plant vigor over a span of time. Some native insects feed on Multiflora Rose, but none are being used as biocontrol agents currently due to a lack of host specificity and effectiveness. Researchers have explored fungal pathogens that may weaken or kill Multiflora Rose, but as of now, there are no widely accepted fungal biocontrol agents. Promoting dense growth of competitive native species might suppress Multiflora Rose through resource competition and allelopathy, though this is more ecological management than direct biological control.

Chemical Control

Chemical control methods can also be effective when attempting to control Multiflora Rose. Glyphosate (Roundup) can be applied as a foliar spray (at a 1-2% solution) during active growth. This is especially effective when plants are flowering or in bud. This method may require repeated applications. Triclopyr (Crossbow or Remedy Ultra) can be used for foliar, basal bark, or cut-stump treatments. It is particularly effective on woody species such as Multiflora Rose. Aminopyralid (Milestone) is a selective herbicide that can be used to control Multiflora Rose. It specifically targets broadleaf weeds and woody plants. It is most effective within pastoral settings because of its minimal impact on native grasses. Imazapyr (Arsenal) is a non-selective herbicide with soil activity. While it can provide long-term control of Multiflora Rose, it may also kill desirable vegetation nearby. Picloram (Tordon) is another chemical control option. It boasts strong residual activity in soil and is effective against Multiflora Rose and other woody invasive species. It does require careful application due to potential leaching and off-target effects. When using any type of herbicides or chemicals, use caution and wear the right PPE to protect yourself from harmful chemicals. Certain chemicals can be harmful for many reasons, including causing burns to the skin or death in some more serious situations. Mechanical, biological, and chemical control practices can be combined in some scenarios to help kill off Multiflora Rose or any other undesired invasive species.

Photos

Fig. 1 This photo was taken from a source (Countryside) showing the white, pinkish flowers that the Multiflora Rose boasts during the growing season. This report was conducted in the late fall when the flowers had already fallen off the plant. The white, pinkish flower shows off the yellow center of the stamen and pistil. If you look at the petals closely, you can see that they have a notch in the center of the petals, resembling the shape of a heart. The flowers are made up of five petals each and grow in clusters/bunches.

Fig. 2 This closeup photo, courtesy of driftless prairies, shows the sharp, curved thorns of Multiflora Rose. These thorns can deter some herbivores from eating the leaves and rosehips on the plant.

Fig. 3 This is an original field photo that shows the distinct bright red rosehips of Multiflora Rose. The rosehips are packed with vitamin C and are safe to eat for humans, animals, and birds. They are commonly used to make tea.

Fig. 4 This is another original field photo that shows the leaves of Multiflora Rose. The leaves are oval shaped with serrated edges.

Concept Map

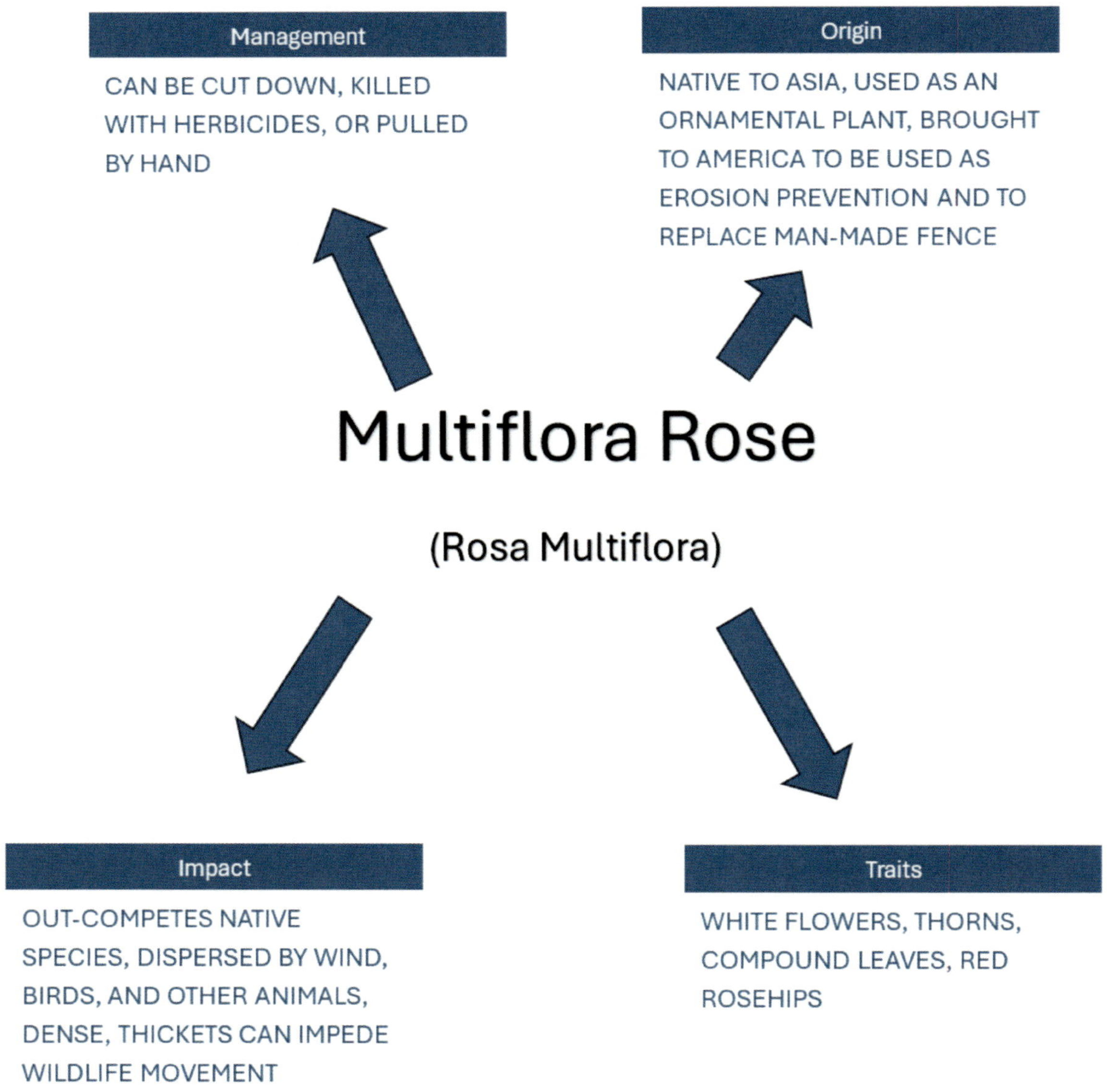

Multiflora Rose was first introduced into the United States in 1866 primarily as an ornamental plant. It gained popularity in the late 1800's and early 1900's as a method of preventing erosion and to replace man-made fencing along field edges. The seeds quickly spread, causing Multiflora Rose to start appearing everywhere, even where it wasn't wanted. The spread can primarily be traced to wind, birds, and animals (among other factors). It can easily be identified due to its white flowers, compound leaves, and bright red rosehips. Its dense thickets and quick spread cause many problems such as impeding natural wildlife movement, polluting fence lines, roadsides, and field edges, and being very hard to control. The plant can be managed by pulling by hand, killed off with herbicides, and/or cut down.

References

A Rose Is a Rose - Driftless Prairies: Native Ecosystems. (2013, July 2). https://driftlessprairies.org/wp-content/uploads/2013/07/IMG_0766-scaled.jpg

Multiflora Rose. Multiflora Rose | National Invasive Species Information Center. (n.d.). https://www.invasivespeciesinfo.gov/terrestrial/plants/multiflora-rose WV. (n.d.-a). https://dep.wv.gov/WWE/getinvolved/sos/SiteAssets/Pages/WVwetflora/MultifloraRose.pdf

Multiflora rose: An invasive species, a cautionary tale, and a good (?) idea gone bad. West Virginia Highlands Conservancy. (2024a, April 12). https://www.wvhighlands.org/article/multiflora-rose-an-invasive-species-a-cautionary-tale-and-a-good-idea-gone-bad/

Top invasive plants - WVDNR. West Virginia Division of Natural Resources. (2023, May 2). https://wvdnr.gov/plants-animals/exotic-and-invasive-species/top-invasive-plants/

Invasion and Control of Tree-of-Heaven (*Ailanthus altissima*)

Kieren Poff

Introduction

The *Ailanthus altissima*, also known as Tree-of-Heaven, is a deciduous tree that has become extremely invasive across the temperate regions of North America and Europe. Due to its invasiveness and striking appearance, it has also adopted the nickname of "alien tree".

This tree can grow up to eighty plus feet tall and can thrive almost anywhere. This plant originated in China and Taiwan and was originally introduced for decorative landscaping purposes in the 1700s. It became popular for its rapid growth rate and its resistance to disease and insects (which are both characteristics that have led to this tree becoming invasive). The tree of heaven has since taken over cleared sites at an unbelievable rate.

This report will show the ecological consequences of *Ailanthus altissima*'s spread, its traits, the factors helping it to spread, and the challenges of trying to control it. It will also look at control and removal methods.

Description

The tree of heaven is known for its striking appearance. It can grow upwards of 80 feet tall. Its unnatural tallness stands out among the native species. It grows 1-2 feet per year which is a much higher rate than native oak species. It has 22-40 leaflets per rachis. Its large leaflets create a dense canopy that prevents sunlight from penetrating the soil to benefit the smaller species of plant. *Ailanthus altissima* has a yellowish-brown colored bark with very prominent and large lenticels, and large leaf scars. It is easy to spot because it is usually found growing in groups. When leaves are crushed it gives off a putrid/rotting smell.

Native Range and Method of Introduction:

Tree of Heaven is native to China and Taiwan.

The first recorded instance of *Ailanthus altissima* being introduced in North America was in 1784 in Philadelphia, Pennsylvania. It was introduced as an exotic ornamental shade tree (Miller, 1990). It later became more wide spread during the Gold Rush through Chinese immigrants who valued the tree for its medicinal properties (FEIS). Nearly every part of the tree has some application in Chinese traditional medicine (iNaturalist). Its spread was also encouraged as a food source for silkworms (Ohio DNR, 2021). By 1902, the U.S. Department of Agriculture had begun distributing the seeds.

Soler & Izquierdo (2024) and Sladonja et al. (2015) states that *Ailanthus altissima* has good and bad sides. Because of its deep roots, the *Ailanthus altissima* helps prevent landslides.

According to the Ecological Landscape Alliance in 2009, the "importation, trade, and sale of tree-of-heaven has been banned" in parts of the United States.

Regions of Invasiveness

Ailanthus altissima has become largely invasive throughout North America, Europe, and Korea, particularly in disturbed sites in urban areas.

A study by Gao et al. (2025) looked at *Ailanthus altissima*'s spread in urban areas. It showed strong relations between disturbed soil and heat pockets. The tree of heaven thrives particularly well in areas cleared for mining and warm climates. The invasiveness of this tree has also become an issue in Europe. A research report by Lee & Lee (2006) showed similar spread in Seoul, where compacted soil and light helped increase the spread.

According to the U.S. Department of Agriculture, the invasive spread is a nationwide issue. In the Eastern and Midwestern US, the tree of heaven has "infested hundreds of acres of oak-hickory forests." The National Invasive Species Information Center cites the *Ailanthus altissima* as being the most invasive species in the South and the West Virginia DNR has listed it as one of the top ten most invasive species in the state.

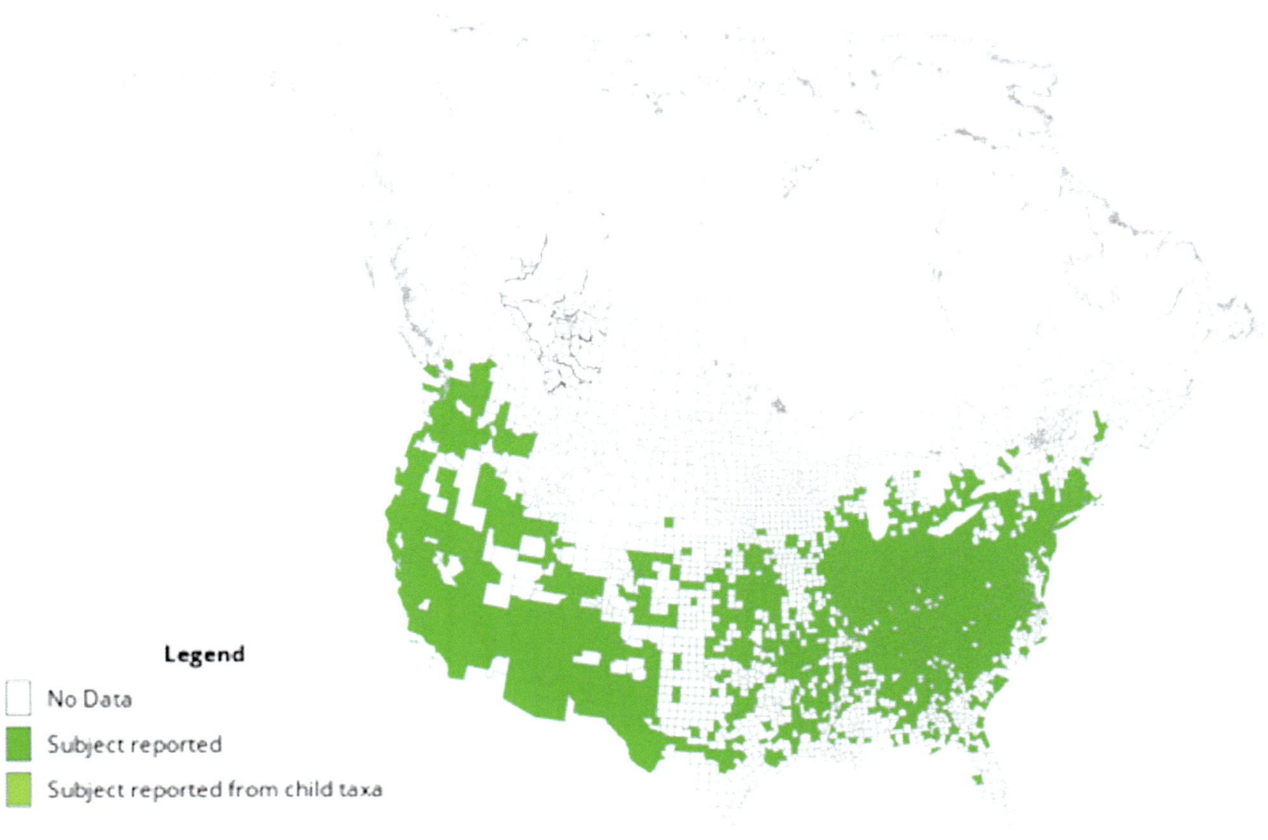

Fig. 1 This map was provided by EDDMaps and shows the distribution of the *Ailanthus altissima* across North America. As you can see, it is less prevalent in the colder regions.

Ecological Impact

The tree of heaven has vast ecological impact. It reproduces rapidly and creates a dense canopy which covers the sunlight on the floor hindering growth to other plants. *Ailanthus altissima* is a root suckering plant, it has aggressive root growth underground which takes the nutrients from the soil. The plant is allelopathic, it secretes a chemical, ailanthone, into the soil that is toxic to other plants, giving itself more room to grow (Sladonja et al. 2015).

The tree of heaven secretes allelopathic compounds which stunt the growth of other plants. These secretions have earned the tree another nickname, the stinking sumac. The chemical secretions are said to smell like rotting flesh or burned peanut butter.

Thompson (2008) described *Ailanthus altissima* as a "highly productive seed-maker", with helicopter-shaped seeds (samaras) that are easily carried by the wind. This tree can reproduce very quickly because it is dioecious, meaning that it reproduces through male and female trees.

When the *Ailanthus altissima* begins to take over an area, this changes the biodiversity of the ecosystem. It changes what plants are available for animals to use as a food source. Native species don't consume the plant because it isn't in their natural diet. In the U.S. the tree of heaven has overtaken oak-hickory forests, decreasing the availability of acorns.

The tree of heaven isn't just an issue for the ecosystem. It is a nuisance to humans as well. The fast-growing intricate root system can damage foundations, sewer systems, and pavement (Nature Conservancy). Due to its massive height and rapid growth, this tree is an issue for overhead powerlines as well.

Control and Management Strategies

Mechanical control cutting is a common attempt, but it can be ineffective due to its quick resprouting ability, sprouts can grow at 10+ feet per year (Isler et al., 2023). Control will require repeated cutting and follow-up.

A recent study by Isler et al. (2023) documented the tree's stubbornness in European forests. Even after mechanical removal, root suckering and seed removal it still showed rapid spread and took over the native plants. Fang et al. (2023) found that the native plants changed the microbes found in the tree of heaven and even encouraged it to grow more.

With chemical control, imazapyr and triclopyr can be effective only when applied to cut stumps. (Lewis & McCarthy, 2008). Hrabovský & Hladík (2024) were successful in slowing the growth using the chemical control process. This method is usually ineffective on sprouts and seed spread.

According to the USDA, herbicide control is a better control method than a prescribed fire because any missed or dormant seedlings are likely to spread rapidly on the freshly burned site.

Due to the hardiness of *Ailanthus altissima* and the fact that it is a root spreader, Lewis & McCarthy (2008) found that using herbicides isn't always an effective way to control growth and spread as it can kill native plants and not harm the tree of heaven in the process. Eco Landscaping recommends applying glyphosate herbicide when the plant is preparing to flower for the most effectiveness due to a higher level of carbohydrates being found in the leaves

A third method is biological control. Research is being done on fungal methods like verticillium nonalfalfae, which could be a possible choice for selective infection (Soler & Izquierdo, 2024). However, this method could be dangerous to nearby native species. According to Fang et al. (2023), eriophyid mites and endophyte manipulation are another possibility.

According to Eco Landscaping, there aren't currently any commercially available insects that are effective enough to be used as a biological control method. As of 2009, *E. chinesis* weevils were being tested as a possible biological control (USDA). However, there are three natural threats to the tree: verticillium wilt, armillaria root rot, and the Cynthia moth larva.

The best method of control is a multifaceted approach; by cutting back trees and injecting the roots with a chemical herbicide, it will have a higher rate of being effective in the long run. This method is called "hack and squirt", a common management practice used by foresters.

Fig. 2 The photo on the left depicts *Ailanthus altissima* growing on a disturbed roadside in Jackson County, West Virginia. It shows the large leaflets and canopy, also the oddly colored bark. The photo on the right shows the oddly colored bark with very prominent and large lenticels as well as the large leaf scars on the bark.

Fig. 3 The photo on the left was taken in Glenville, West Virginia. This tree is growing on a hillside near Glenville State University's public safety office. This tree is a great representation of how other species don't grow close to the tree due to its allelopathic capabilities. This photo shows the high growth rate as this tree is less than 15 years old. The photo on the right, taken by John M. Randall of The Nature Conservancy, shows the understory growth of *Ailanthus altissima*.

Fig. 4 This photo, taken by John Woodmansee of the Purdue University Extension Office, shows the yellowish seed clusters on the tree of heaven. These clusters only grow on the female trees

Fig. 5 This photo, taken by John Woodmansee of the Purdue University Extension Office, shows the glandular teeth at the base of the leaflets.

Concept Map

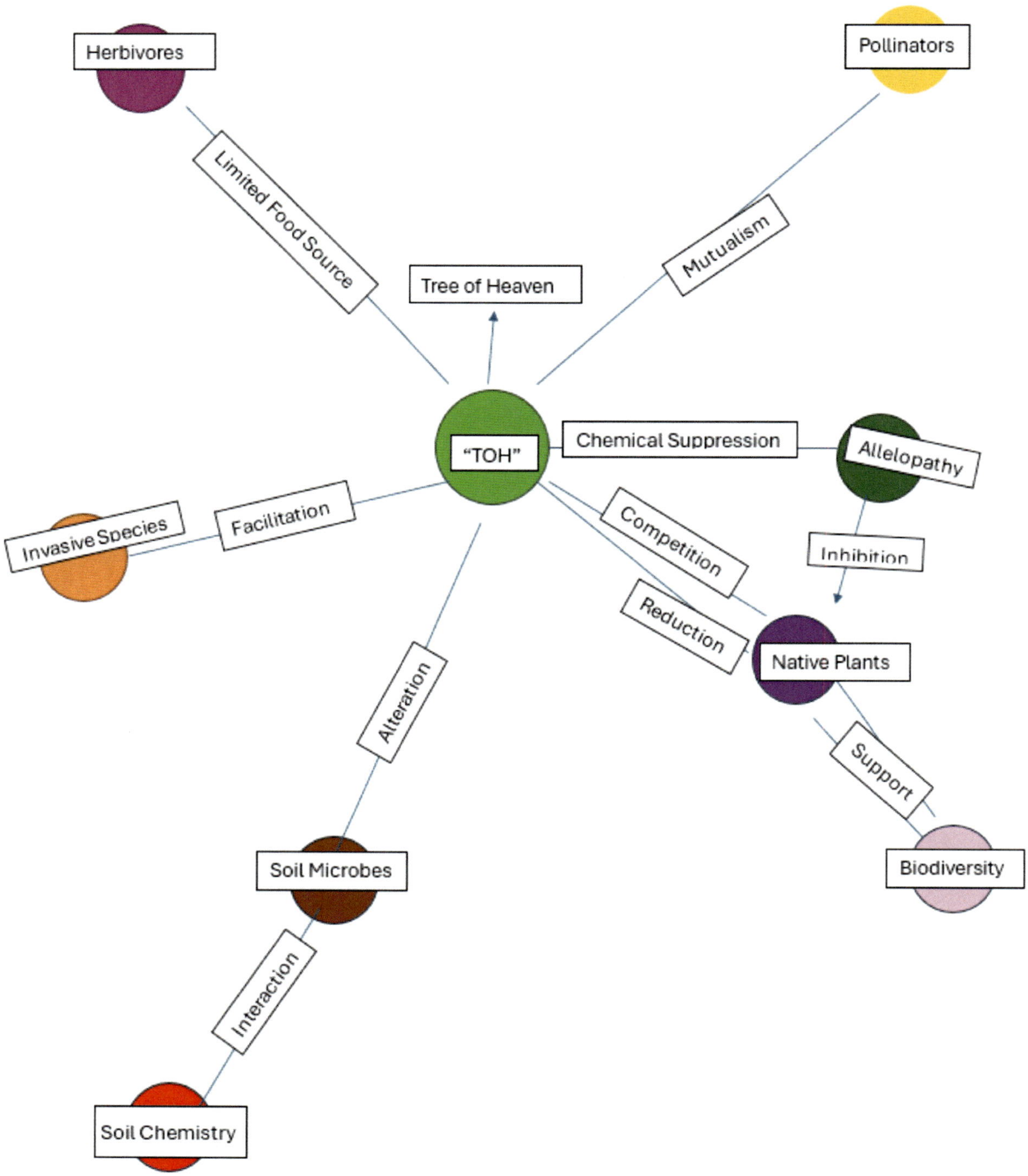

Explanation of the Concept Map

Allelopathy: It releases toxins like ailanthone into the soil, which slows down growth of native plants. This chemical disrupts the plants native habitat and highly disrupts the plants in the native range.

Soil Microbes: Its leaf litter and root secretions alter microbes, reducing symbiotic fungi that live on the roots of the plants and help prevent root rot. Without these fungi, plants in wetter regions will suffer.

Pollinators: Though it provides nectar, it overshadows native flowering plants and reduces diversity.

Herbivores: Because of its strong odor and poor taste, animals don't feed on it. Herbivores enjoy eating native plants instead. This increases pressure on the already struggling native species and leads to over browsing.

Tree of Heaven often teams up with other invasive species like the spotted lanternfly, which uses it as a host. This teamwork sets the stage for maximum invasive success. The spotted lantern fly does harm this tree slightly by eating it, but it is not enough to kill it.

In regions like Appalachia, Tree of Heaven thrives in disturbed areas such as roadsides, urban lots, and forest edges. Its rapid growth and hardiness it a pioneer species. Its spread leads to loss of native understory plants which reduces biodiversity, reduces habitat variety for native wildlife, and makes it difficult to regrow forests.

Conclusion

The tree of heaven has taken a toll on the ecosystem of North America and the world. Its ecological impact has disturbed many ecosystems and has had a catastrophic effect on the native species' ability to thrive in their natural habitats.

Ailanthus altissima is one example of how complex invasive species management can be. Due to its ability to adapt and reproduce at alarming rates it is a very difficult tree to manage and control. In knowing how invasive the tree of heaven is, we can better educate the public on this invasive species and stop the spread.

References

A Tale of Two Invaders: Tree of Heaven and Spotted Lanternfly. (n.d.). Integrated Pest Management. https://www.canr.msu.edu/resources/a-tale-of-two-invaders-tree-of-heaven-and-spotted-lanternfly

Ailanthus altissima. (n.d.). Www.fs.usda.gov. https://www.fs.usda.gov/database/feis/plants/tree/ailalt/all.html

Burns, R. M., Honkala, B. H., & Technical coordinators. (2022). *Silvics of North America: Volume 2. Hardwoods*. United States Department of Agriculture (USDA), Forest Service, Agriculture Handbook 654. https://research.fs.usda.gov/treesearch/1548

EDDMapS. (2025). EDDMapS.org. https://www.eddmaps.org.

Fang, K., Yang, A.-L., Li, Y.-X., Zeng, Z.-Y., Wang, R.-F., Li, T., Zhao, Z.-W., & Zhang, H.-B. (2023). Native plants change the endophyte assembly and growth of an invasive plant in response to climatic factors. *Applied and Environmental Microbiology*, *89*(10). https://doi.org/10.1128/aem.01093-23

Gao, R., Song, Z., Zhao, J., & Li, Y. (2025).). Predicting the distribution of Ailanthus altissima using deep learning-based analysis of satellite imagery. https://doi.org/10.3390/sym17030324

Hendrix, J. (2024, July 23). *Double trouble: tree of heaven and spotted lanternfly threaten our environment - Olmsted Parks Conservancy*. Olmsted Parks Conservancy. https://www.olmstedparks.org/double-trouble-tree-of-heaven-and-spotted-lanternfly-threaten-our-environment/

Hrabovský, M., & Marko Hladík. (2024). Control of invasive Ailanthus altissima in the Danube floodplain forests in Bratislava using chemical and biological agents. *Forest Ecology and Management*, *569*, 122225–122225. https://doi.org/10.1016/j.foreco.2024.122225

INaturalist. Tree of heaven (Plants of Saxony) · iNaturalist. (n.d.). INaturalist. https://www.inaturalist.org/guide_taxa/635939

Invasive to Avoid: Tree-of-Heaven. (n.d.). Wildlife.ca.gov. https://wildlife.ca.gov/Conservation/Plants/Dont-Plant-Me/Tree-of-Heaven

Invasive species centre. (n.d.). *Tree of heaven*. Invasive Species Centre. https://www.invasivespeciescentre.ca/invasive-species/meet-the-species/invasive-plants/tree-of-heaven/

Isler, J., Harald Bugmann, Conedera, M., & Thrippleton, T. (2023). Long-term dynamics of tree of heaven (Ailanthus altissima) in central European forests. *European Journal of Forest Research*, *142*(5), 1149–1166. https://doi.org/10.1007/s10342-023-01582-9

Lee, H., & Lee, C. (2006). Environmental factors affecting establishment and expansion of the invasive alien species of tree of heaven (Ailanthus altissima) in Seoripool Park, Seoul. *Integrative Biosciences*, *10*(1), 27–40. https://doi.org/10.1080/17386357.2006.9647281

Lewis, P. (2014, May 15). Tree of Heaven: An Exotic Invasive Plant Fact Sheet - Ecological Landscape Alliance. Ecological Landscape Alliance. https://www.ecolandscaping.org/05/landscape-challenges/invasive-plants/tree-of-heaven-an-exotic-invasive-plant-fact-sheet/

Lewis, K., & McCarthy, B. (2008). Nontarget Tree Mortality after Tree-of-Heaven (Ailanthus altissima) Injection with Imazapyr. *Northern Journal of Applied Forestry*, *25*(2), 66–72. https://doi.org/10.1093/njaf/25.2.66

Proceedings, 20th U.S. Department of Agriculture Interagency Research Forum on Gypsy Moth and Other Invasive Species, 2009. (2025). Google Books. https://books.google.com/books?id=-qTJI7RE38IC

Randall, J. M. (n.d.). [Photograph]. The Nature Conservancy.

Sladonja, B., Sušek, M., & Guillermic, J. (2015). Review on Invasive Tree of Heaven (Ailanthus altissima (Mill.) Swingle) Conflicting Values: Assessment of Its Ecosystem Services and Potential Biological Threat. *Environmental Management*, *56*(4), 1009–1034. https://doi.org/10.1007/s00267-015-0546-5

Soler, J., & Izquierdo, J. (2024). The Invasive Ailanthus altissima: A Biology, Ecology, and Control Review. *Plants*, *13*(7), 931. https://doi.org/10.3390/plants13070931

The Nature Conservancy. (2020, July 6). *Journey with Nature: Tree of Heaven*. The Nature Conservancy. https://www.nature.org/en-us/about-us/where-we-work/united-states/indiana/stories-in-indiana/journey-with-nature--tree-of-heaven/

Thompson, J., Fell, R., Brewster, C., Lloyd, P., & Stipes, H. (2008). *Pollination Biology of Ailanthus altissima (Mill.) Swingle (Tree-of-Heaven) in the Mid- Atlantic United States*. https://vtechworks.lib.vt.edu/bitstream/handle/10919/32586/Thompson_Thesis_Final_5-31-08.pdf

Tree-of-Heaven. (2021). Ohiodnr.gov. https://ohiodnr.gov/discover-and-learn/plants-trees/broad-leaf-trees/tree-of-heaven-ailanthus-altissima

Tree of Heaven Serves as Devilish Host to Spotted Lanternfly : USDA ARS. (2025). Usda.gov. https://www.ars.usda.gov/oc/dof/tree-of-heaven-serves-as-devilish-host-to-spotted-lanternfly/

U.S. Department of Agriculture, Forest Service. (2025). Ailanthus altissima: Fire effects information system. U.S. Forest Service. https://www.fs.usda.gov/database/feis/plants/tree/ailalt/all.html

Wisniewski, N. (2025, August 26). *The Invasive Tree You May Not Want To Get Rid Of If You Have A Spotted Lanternfly Problem*. Outdoor Guide. https://www.outdoorguide.com/1946890/invasive-tree-of-heaven-get-rid-trap-spotted-laternflies/

Woodmansee, J. (2023, August). Identifying tree-of-heaven [Photograph]. Purdue University Extension. https://extension.purdue.edu/news/county/whitley/2023/08/identifying-tree-of-heaven.html

B

Southern Luzon State University

Invasive Species *Imperata cylindrica* (Cogon Grass): An Ecological and Economic Threat

Bernadette T. David, Blisslyn V. Nañola & Edralin P. Angara

Species History, Information, and Characteristics

Imperata cylindrica, known commonly as Cogon Grass, is native to a broad area that spans Southeast Asia, eastern and southern Africa, and Oceania. Historically, the grass was distributed to Japan, China, India, Thailand, Malaysia, Indonesia, the Philippines, parts of Australia, and parts of Africa. The perennial grass flourishes in a tropical and subtropical climate (Holm et al., 1977; Daneshgar et al., 2008; Patterson & Flint, 2001). The distribution of Cogon grass to areas outside its native range is primarily the result of human activities, and its arrival dates back as early as the 19th century (Daneshgar et al., 2008). The following is a comprehensive list of ways the grass was likely introduced and transported to new areas, listed in order of importance in the spread of the grass:

Cogon grass has become invasive in many regions of the world due to both incidental and deliberate human actions. According to Daneshgar et al. (2008), one of the most important historical pathways to establish was the contamination of crop seed during the expansion of agriculture. As logistics systems expanded to support agriculture, cogon grass (and its seed) was carried across long distances by accident when providing agricultural supplies to new sites, where they subsequently grew and became established as a coexisting crop. Another method of accidental movement through human activity occurred with the international trade of goods from Asia. Daneshgar et al. (2008) describe cogon grass as a commonly used packing material for shipping goods to other countries, which often ended up discarded when the goods were unpacked upon arrival at the destination. This resulted in the unintentional release of cogon grass and its rooting in many new invasive sites.

The species of Cogon grass was introduced to an area accidentally and intentionally as a result of poor agricultural management practices. The plant was introduced mainly to control soil erosion, stabilize soil, and serve as forage for livestock. Trautwig et al. (2017) indicate that these types of introductions, such as when Cogon grass was introduced to the Southeast United States for livestock trials in the 20th century, only hurt farms and ranches, as it provided no edible forage and was quickly transformed into an invasive species. There were also aesthetic reasons that farmers and ranchers introduced the plant; Daneshgar et al. (2008) indicate that many people did not realize the potential for the plant to become an aggressive invasive species and introduced it strictly as an ornamental plant.

Cogon grass populations are still expanding because of modern-day construction and logistical activities. The ability of cogon to survive is demonstrated through the ease with which it can be moved by mechanical means using rhizomes (a type of root that can also grow into new plants). According to Daneshgar et al. (2008), the movement of heavy machinery, vehicles, and contaminated soil with rhizomes has caused the continued invasion of cogon grass because it can transport itself via machines from previously pest-free areas and has established new means for it to invade from seed (historically) to via contemporary industrial means.

Ecological and Economic Implications

Many people know that cogon grass is a highly invasive species that can alter how ecosystems are formed and function greatly (Russell, 2017; Daneshgar et al., 2008). One of the principal factors driving this type of ecological disruption is a decreased amount of biodiversity. As pointed out by Daneshgar et al. (2008) and Patterson and Flint (2001), cogon grass can easily establish itself as a dense, continuous monoculture throughout many different types of ecosystems. Cogon grass competes with many native plants for essential resources like light, water, and nutrients, lowering the biodiversity of the area and harming the surrounding health of the ecosystem. Any additional allelopathic effects will also contribute to the successful competition by cogon. Some research (Kato-Noguchi, 2022; Tokunaga and Imura, 2019) has shown that the chemicals produced by cogon are capable of suppressing the germination and growth of the surrounding plants, which provides cogon with greater competition and helps it dominate the surrounding plant community.

Aside from affecting its surroundings, cogon grass's physical properties alter naturally occurring fire patterns. The high amounts of oil contained in the leaves and stems and the large amount of dry biomass accumulated by cogon grass make it one of the most capable tinder materials in the area, especially in drier seasons (Stewart & Finlayson, 2021; Daneshgar et al., 2008). This ability to burn creates conditions that increase the frequency, intensity, and severity of wildfires. By creating lethal conditions for non-adapted native species during a wildland fire, cogon grass can quickly re-establish itself in the burnt-over land from the protected rhizomes (Stewart & Finlayson, 2021) and significantly degrade previously existing habitats due to increased plant density and altered fire regimes (Stewart & Finlayson, 2021). Degraded habitats support smaller numbers of birds, mammals, and insects that rely on the diversity available within native plant communities to fulfill their food and shelter needs.

Based on Daneshgar et al., 2008 economic impacts of cogon grass report, it has an environmental effect as well as creating significant agricultural, forest, and land management economic challenges. In the agricultural industry, cogon grass invades farmland and reduces crop yields from sources like corn, soybean, and citrus by directly competing for resources with them. Farmers also incur high continuing costs for treatments to control cogon grass (Breeyen 2007, Daneshgar et al., 2008). In forestry, reforestation is also disrupted as cogon grass disrupts seedling growth and establishment, leading to lower timber production and greatly increased reforestation costs in comparison with non-invaded areas (MacDonald et al., 2018).

Economic pressure impacts the management of both livestock and infrastructure. Cogon grass is a plant that is not of good quality for grazing; however, when it grows in an area where other natural species are already grown, it will replace those palatable native species, thus reducing the overall carrying capacity of that area for the grazing of livestock. Because of this loss of available grazing land, farmers must increase their expenditures on feed or new livestock to replace those that cannot be supported by the area. Additionally, because of the aggressive growth of cogon grass, it will rapidly invade the area surrounding the roads, utility easements, and recreational easements; therefore, clearing or maintaining these will be very costly. Finally, the increase in fire danger creates an enormous financial burden for farmers because the cost of fire prevention, suppression, and recovery will increase (Daneshgar et al., 2008; Stewart & Finlayson, 2021).

Control and Eradication Measures

Management of cogon grass can often be difficult due to its extensive rhizome system, high reproductive potential (both asexual and sexual), and negative response to most management practices (Gholami et al., 2022; Daneshgar et al., 2008; USDA Forest Service, 2015). In most cases, a more comprehensive, integrated-control approach may be necessary to complete the management and elimination process.

Mechanical Management

Regular cutting or mowing has the potential to reduce above-ground biomass and temporarily suppress growth. However, Gholami et al. (2022) emphasize that it needs to take place on a regular basis for several years to lower rhizome energy reserves. For the best results, use this method along with other control methods. Deep plowing or disking can destroy rhizome systems, but Daneshgar et al. (2008) caution that if not deep enough, there is a risk of fragmenting rhizomes, which can lead to the infestation spreading to new sites. This method is frequently utilized in an agricultural setting before replanting. Hand pulling and excavation are only practical for small, very recent infestations; as per the USDA Forest Service (2015), it is important to remove the entire rhizome system, or else reestablishment will occur. It is also very costly and labor-intensive.

Biological Control

For Cogon grass, biological control agents are mostly still in the research phase (Gholami et al., 2022). Van Driesche (2000) notes that some agents have been found to have potential as biological control agents (e.g., rust fungi, such as *Puccinia imperatae*; various insects), but host specificity and potential impacts on native, endemic grass species have posed challenges to obtaining significant approval for use. A moderate amount of research is still ongoing to evaluate highly host-specific pathogens or herbivores that can attack Cogon Grass and minimize negative impacts on economically important commercial crops or potentially threaten native biodiversity (Gholami et al., 2022).

Chemical Control

Glyphosate-based herbicides are usually the most effective means of chemical control for Cogon Grass (MacDonald et al., 2018). Breeyen (2007) found that most cases require repeated applications (usually annually) over multiple years to achieve sufficient control or eradication. Imazapyr is another herbicide that can be used in specific contexts, especially forestry, but has residual effects in soil industries of concern (USDA Forest Service, 2015). Herbicides tend to work best when applied to actively growing grass. Gholami et al. (2022) indicate that cogon grass has a summer-to-fall growth period during which nutrients are transported to its rhizomes, which is the best time for control. A combination of glyphosate and imazapyr is generally more effective than using either product individually, especially for dense infestations (Gholami et al., 2022). Burning followed by herbicide application: Prescribed burning can reduce the aboveground biomass before an herbicide application to facilitate fresh germination regrowth. Prescribed burning should be carefully managed to avoid wildfires (USDA Forest Service, 2015).

Original Field Photos

Fig. 1 Dense monoculture of Cogon grass invading an abandoned field in a rural area in Lucena City, Quezon, Philippines.

Fig. 2 White, fluffy seed head inflorescence of Cogon grass

Fig. 3 Leaf blade of Cogon grass

Concept Map

Cogon grass is a rapidly growing, invasive perennial species that primarily occurs in tropical and subtropical areas. It has a significant effect on ecosystems because of its interactions with living and abiotic components, as shown in Fig. 4.

From an abiotic perspective, cogon grass thrives in disturbed soils with little nutrient availability, thus changing the chemistry of the soil through its effect on silica and organic matter (Daneshgar et al., 2008). Stewart and Finlayson (2021) point out its relatively high flammability, which increases the frequency and intensity of wildfires. Ironically, the grasses then swiftly grow back in these areas that had burned. It does prefer sunny, moist conditions and can densely mat other vegetation to prevent moisture infiltration to competing vegetation (Daneshgar et al., 2008).

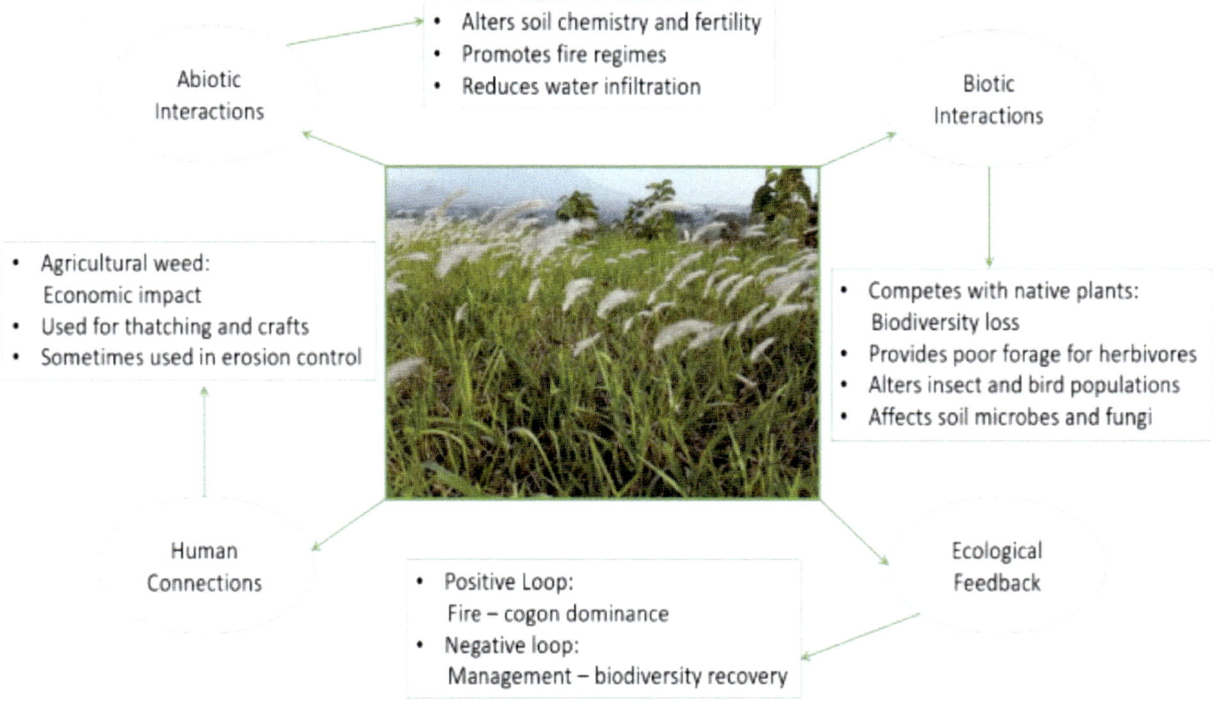

Fig. 4 Interconnected Interactions, Ecological Feedback, and Human Connections of Cogon grass

From a biotic perspective, it intensely competes for space with native flora, creating dense rhizomes and using allelopathic chemicals that inhibit growth and diminish the biodiversity of plants (Daneshgar et al., 2008; Kato-Noguchi, 2022). It is very low in nutrients for grazing fauna like cattle, and while it does provide some food for a limited number of insects, insect biodiversity is negatively affected. It provides nesting cover for some birds and small mammals, but it has limited native food sources, thus changing where faunal communities, predators and prey can interact. Daneshgar et al. (2008) further elaborate that it alters soil pH and nutrient cycling, ultimately altering microorganisms and fungi living in the soil, thus disrupting critical symbiotic relationships. Cogon grass is an agricultural adversary, reducing agricultural production and increasing costs at the human-ecological interface (Daneshgar et al., 2004). Although cogon grass can inadvertently cause soil erosion, its potential to disrupt ecological systems and processes poses a greater concern. Yet, there are cultural uses in different areas, for example, as a roof thatch and for crafts.

The recognition of positive and negative feedback loops for this species is identifiable: Stewart and Finlayson (2021) and Gholami et al. (2022) describe how cogon grass creates a positive feedback loop by facilitating intense fires that kill native plants, clear space for further cogon growth, and allow cogon grass to continue its growth cycle. There is also a potential negative feedback loop that can be initiated by effectively controlling and managing cogon grass with the goal of converting formerly infested sites back to native plant cover, biodiversity, and reduced cogon grass spread (Gholami et al., 2022).

References

Breeyen, A. D. (2007). *Biological control of Imperata cylindrica in West Africa using funga pathogens.* https://ufdcimages.uflib.ufl.edu/UF/E0/02/17/64/00001/denbreeyen_a.pdf.

Daneshgar, P., et al. (2008). *Cogon grass (Imperata cylindrica), an Alien Invasive Grass, Reduces Survival and Productivity of an Establishing Pine Forest. Forest Science, 54*(6), 579–587.

Gholami, H., et al. (2022*). Integrated Management of Imperata cylindrica: A Review of Sustainable Control Strategies. Journal of Agricultural Science and Technology, 24*(1), 1- 15.

Holm, L.G., et al. (1977). The World's Worst Weeds: Distribution and Biology. University Press of Hawaii.

Kato-Noguchi H. *Allelopathy and Allelochemicals of Imperata cylindrica as an Invasive Plant Species. Plants (Basel). 2022 Sep 28;11*(19):2551. doi: 10.3390/plants11192551. PMID: 36235415; PMCID: PMC9573136.

MacDonald, G.E., Johnson, C.M., Minogue, P.J., & Lauer, D.K. (2018). *Cogon Grass (Imperata cylindrica) Management in Florida Forest Lands*. University of Florida IFAS Extension. Publication #FOR194.

Patterson, D.T., & Flint, E.A. (2001). *Ecophysiology of Imperata cylindrica: A Global Perspective. Biological Invasions, 3*(3), 223-239.

Russell, J. C., et al. (2017*). Invasive alien species on islands: impacts, distribution, interactions, and management. Environmental Conservation, 44*(4), 359–370.

Stewart, B.D., & Finlayson, W.B. (2021). *Fire Ecology of Invasive Grasses: Case Study of Imperata cylindrica. Environmental Science & Policy, 115,* 120-130.

Tokunaga, S., & Imura, O. (2019). *Allelopathic Effects of Cogon Grass (Imperata cylindrica) on Native Plant Germination and Growth. Plant Ecology and Evolution, 152*(1), 77-85.

Trautwig, A. N., et al. (2017). *Cogon grass (Imperata cylindrica) Affects Above- and Belowground Processes in Commercial Loblolly Pine (Pinus taeda) Stands. Forest Science, 63*(1), 10–16.

USDA Forest Service. (2015). *Field Guide for the Identification and Management of Invasive Plants in Southern Forests*. Southern Research Station. (Government publication, highly credible for regional management)

Van Driesche, R.G., & Van Driesche, S. (2000). *Biological Control of Invasive Plants in the Eastern United States.* USDA Forest Service. Publication FHTET-2000-04. (Relevant for biological control efforts)

Lantana camara: Spread, Impacts, and Control

Ronald Kristoffer S. Gayeta, Bernadette D. Mancenido & Wilson C. Ramos

Species History, Range, and Characteristics

Invasive species could be described as destructive, outcompeting, and harmful. United States Geological Survey (USGS) defines them as introduced, non-native organism that could be in the form of a disease, parasite, plant, or animal that has the capacity of proliferating at rapid rates from the site of its original introduction and has the potential of causing harm not only to the environment, but to society as well, particularly to human health and the economy. Given the widespread effect on different sectors of this environmental issue, it necessitates this study, which entails vital information regarding its origins, proliferation, and control measures that would enable communities to manage and prevent further damage to biodiversity and society.

A native to Central and South America, including Mexico, the Caribbean, and tropical parts of America, the West Indian Lantana, scientifically known as *Lantana camara*, has an original range extending from Mexico through Central America to northern South American countries such as Venezuela and Colombia. It is a plant that thrives in both tropical and subtropical climates, adaptable across a variety of ecosystems within the native range (Fern, 2025). It was introduced in the Philippines from Hawaii as part of an exchange program between the United States and the Philippines, initially brought for ornamental purposes but escaped cultivation and has successfully naturalized across the islands (iNaturalist, 2016). This highly adaptive species underwent extensive selective breeding throughout the 17th and 18th centuries for use as an ornamental plant, which resulted in different forms of the species throughout the world (Global Invasive Species Database, 2025).

Lantana is a perennial, upright, low or sub-scandent shrub that grows around two meters in height and forms dense thickets, often monospecific stands that outcompete native vegetation and alter ecosystem structure (Mishra et al., 2023). It has strong recurved thorns and fruits with a strong odor. It has ovate, opposite, hairy leaves with serrated margins that are rough on their surface and scabrous on both sides. The species produces showy, multicolored tubular flowers in compact clusters that may vary from yellow and orange to pink or red and often change color as they age, a pattern linked to pollinator attraction (PMC, 2023). Flowering occurs around April and May (Invasoras pt, n.d.). Following pollination, these inflorescences develop into small, glossy, spherical drupes that transition from green to purplish-black upon ripening and are readily dispersed by birds and other animals (Mridula et al., 2025). Once this shrub reaches sexual maturation around 2-3 years, it reproduces sexually and produces hundreds to thousands of seeds per plant per year. It is dispersed naturally through endozoochory, favored by planting, abandoned lands, or the alteration of an area (Invasoras pt., n.d.). *L. camara* also exhibits rapid growth, high reproductive output, and allelopathic effects on surrounding vegetation, suppressing native plant germination and recruitment, which contributes to reduced biodiversity and altered habitat function in invaded areas (Mdpi Plants, 2021).

Control Measures and Management

The invasion of Lantana has emerged as a significant ecological and agricultural issue in tropical and subtropical areas, including the Philippines. As an invasive species, it displaces indigenous vegetation, modifies soil chemistry, and disrupts forest regeneration. Recent research underscores that no singular approach will completely eradicate *L. camara*; rather, a combination of management strategies is essential for sustained suppression and ecological restoration (Day et al., 2003; Mridula et al., 2025).

Mechanical Control

Mechanical control represents the most direct and immediate approach for eradicating *L. camara*. This method entails the complete extraction of the plant, roots included, to inhibit regeneration. Kalaichelvi (2020) claims that the most efficacious mechanical technique involves uprooting the plant beneath the crown bud, retaining merely two to three centimeters of the root in the soil. This guarantees the entire removal of the regenerative buds, which usually induce resprouting. The research indicated that plants uprooted in this fashion showed no regrowth for as long as three months, particularly when conducted during the rainy season, when the soil is moist and more manageable. Mechanical clearing may encompass cutting and burning, especially in cases of dense infestations. Day, Wiley, Playford, and Zalucki (2003) showed that cutting lantana near the base, succeeded by burning or drying, significantly diminishes above-ground biomass. Nevertheless, this must invariably be succeeded by the replanting of competing native species, as exposed soils can promote the swift re-establishment of lantana. Consistent cutting every three to six months is essential to eradicate seedlings arising from lantana's enduring seed bank, which can remain viable for as long as five years (Mridula et al., 2025). In extensive infestations, heavy machinery like bulldozers and brush cutters may be employed; however, soil disturbance must be restricted to avert reinvasion.

Chemical Control

Chemical control approaches augment mechanical clearance and are very efficacious in inhibiting regrowth. Herbicides, including glyphosate, triclopyr, and 2,4-D sodium salt, have been extensively employed to eliminate *L. camara*. The cut-stump approach is highly effective: lantana stems are severed near the ground, and herbicide is promptly applied to the freshly cut surface, facilitating chemical absorption into the vascular tissue to eradicate the root system. Kalaichelvi (2020) revealed that the application of a 2,4-D sodium salt paste combined with glyphosate wipe post-cutting effectively eradicated lantana in multiple hilly regions, with no signs of regrowth seen.

Day et al. (2003) showed that basal bark treatment, involving the application of a triclopyr and diesel mixture to the bottom 30 cm of the stem, efficiently eradicates woody lantana plants. Simultaneously, foliar application of 1–2% glyphosate is appropriate for young regrowth, dependent upon the plants being in active growth and not experiencing drought stress. The most efficacious outcomes are attained when chemical management is integrated into a sequential strategy—commencing with cutting, followed by herbicide application, and concluding with the replanting of indigenous flora (Mridula et al., 2025). Nonetheless, dependence solely on herbicides may pose environmental hazards; hence, the integration of mechanical and biological approaches is advocated.

Biological Control

Biological control offers a sustainable, long-term strategy for diminishing Lantana populations by the introduction or augmentation of natural predators that target the plant. Over 30 biocontrol agents have been examined worldwide, although only a limited number have demonstrated consistent efficacy. The lantana insect (*Teleonemia scrupulosa*), leaf-mining beetle (*Uroplata girardi*), seed fly (*Ophiomyia lantanae*), and leaf miner (*Octotoma scabripennis*) are identified as the most efficacious (Day et al., 2003; Broughton & Hannan-Jones, 2003). These insects diminish the plant by consuming its leaves, flowers, and seeds, so impairing its photosynthetic efficiency and reproductive success. Gradually, this biological suppression enables native plants to re-establish and compete more efficiently. Nonetheless, biological control rarely results in total eradication because of factors including climatic fluctuations, hybridization among lantana species, and the restricted host range of the agents (Broughton & Hannan-Jones, 2003). Therefore, biological control should be regarded as a supplementary strategy—offering continuous suppression while mechanical and chemical approaches accomplish an initial decrease.

An integrated weed management (IWM) is sought to be the most successful method for its eradication. Its success stems from the integration of mechanical, chemical, and biological controls that are implemented in a systematic order. The structure of this system includes Preliminary clearance (mechanical), which mechanically eradicates adult lantana specimens and disposes of the biomass through incineration. Another component of this structure is targeted chemical treatment, where herbicide is administered to freshly severed stumps or regrowth within 24 hours, absorbed through the leaves or cut system, moving through the plant's vascular system and killing the roots and shoots, evidently preventing regrowth from the root system. In terms of biological controls, biological suppres-sion and revegetation are explored. Biological suppression enables biocontrol agents such as insects or pathogens that exert sustained pressure on the targeted invasive species, lowering their competitive advantage against native species. Revegetation, on the other hand, introduces rapidly growing indigenous trees or grasses to surpass lantana seedlings. Overarching all these controls is monitoring, whereas follow-ups on clearing and herbicide application are done every six to twelve months, to ensure the management's effectiveness.

Day et al. (2003) indicated that effective implementation of integrated management can diminish lantana coverage by more than 70% during a span of two to three years. Kalaichelvi (2020) and Mridula et al. (2025) underscored that the combination of uprooting, chemical eradication, and revegetation led to total elimination across various field sites. In the Philippine context, analogous methods are employed in DENR-led forestry initiatives, when Lantana is either mechanically or chemic-ally eradicated before the introduction of native species. The literature indicates that the eradication of *L. camara* necessitates both prompt removal and sustained suppression efforts. Mechanical control facilitates immediate elimination, chemical approaches inhibit regrowth, and biological treatments provide sustained population reduction over time. When integrated through weed control, these strategies constitute a complete and sustainable approach to diminishing lantana numbers and rehabilitating natural habitats.

Original Field Photos

Fig. 1 Field photographs of Lantana (*L. camara*): A. Growth habit showing a woody shrub with young fleshy stems and mature woody stems, B. Small, multicolored flowers arranged in compact, rounded inflorescences, C. Oval-shaped leaves with serrated margins and a rough, hairy surface, D. Small, spherical berries that change in color from green to black upon maturity

Concept Map

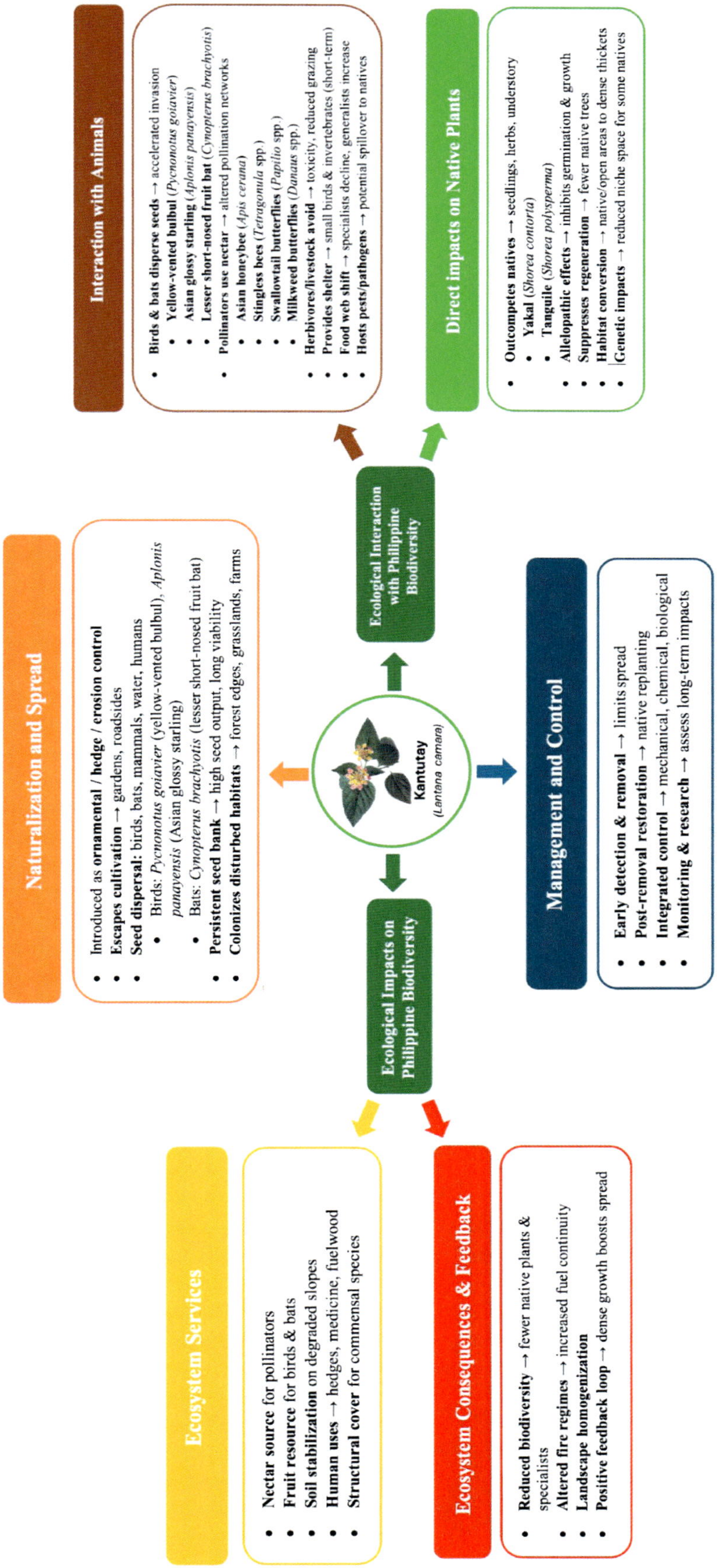
Kantutay
(Lantana camara)
Naturalization and Spread
Introduced as ornamental / hedge / erosion control
Escapes cultivation → gardens, roadsides
Seed dispersal: birds, bats, mammals, water, humans
Birds: Pycnonotus goiavier (yellow-vented bulbul), Aplonis panayensis (Asian glossy starling)
Bats: Cynopterus brachyotis (lesser short-nosed fruit bat)
Persistent seed bank → high seed output, long viability
Colonizes disturbed habitats → forest edges, grasslands, farms
Ecological Interaction with Philippine Biodiversity
Interaction with Animals
Birds & bats disperse seeds → accelerated invasion
Yellow-vented bulbul (Pycnonotus goiavier)
Asian glossy starling (Aplonis panayensis)
Lesser short-nosed fruit bat (Cynopterus brachyotis)
Pollinators use nectar → altered pollination networks
Asian honeybee (Apis cerana)
Stingless bees (Tetragonula spp.)
Swallowtail butterflies (Papilio spp.)
Milkweed butterflies (Danaus spp.)
Herbivores/livestock avoid → toxicity, reduced grazing
Provides shelter → small birds & invertebrates (short-term)
Food web shift → specialists decline, generalists increase
Hosts pests/pathogens → potential spillover to natives
Direct impacts on Native Plants
Outcompetes natives → seedlings, herbs, understory
Yakal (Shorea contorta)
Tanguile (Shorea polysperma)
Allelopathic effects → inhibits germination & growth
Suppresses regeneration → fewer native trees
Habitat conversion → native/open areas to dense thickets
Genetic impacts → reduced niche space for some natives
Management and Control
Early detection & removal → limits spread
Post-removal restoration → native replanting
Integrated control → mechanical, chemical, biological
Monitoring & research → assess long-term impacts
Ecological Impacts on Philippine Biodiversity
Ecosystem Services
Nectar source for pollinators
Fruit resource for birds & bats
Soil stabilization on degraded slopes
Human uses → hedges, medicine, fuelwood
Structural cover for commensal species
Ecosystem Consequences & Feedback
Reduced biodiversity → fewer native plants & specialists
Altered fire regimes → increased fuel continuity
Landscape homogenization
Positive feedback loop → dense growth boosts spread

The concept map of *L. camara* invasiveness is shown on the previous page. Lantana has become widely naturalized in the Philippines after its intentional introduction as an ornamental plant, hedge, and erosion-control species. From cultivated settings, it escaped into gardens, roadsides, farms, and other disturbed areas (GISD, 2023). Its spread is strongly facilitated by seed dispersal through frugivorous animals such as the yellow-vented bulbul (*Pycnonotus goiavier*), Asian glossy starling (*Aplonis panayensis*), and the lesser short-nosed fruit bat (*Cynopterus brachyotis*), as well as by water runoff and human transport. The species produces large quantities of seeds that remain viable in the soil for long periods, forming a persistent seed bank that allows rapid re-establishment and colonization of disturbed habitats, including forest edges, grasslands, abandoned farms, and roadside environments (Taylor et al., 2012).

Once established, *L. camara* negatively affects native plant communities by aggressively outcompeting seedlings, herbs, and understory species through rapid growth and dense thicket formation. Native tree species such as Yakal (*Shorea contorta*) and Tanguile (*Shorea polysperma*), as well as pioneer species like Binunga (*Macaranga tanarius*) and Anabiong (*Trema orientalis*), experience reduced seedling survival under lantana cover. In addition, allelopathic chemicals released from lantana leaves and litter inhibit the germination and growth of understory plants such as Malatungaw (*Melastoma malabath-ricum*), suppressing forest regeneration and altering plant community composition (Sharma et al., 2005). Over time, this leads to habitat conversion from diverse native vegetation to structurally simplified lantana-dominated thickets, reducing ecological niche space for native flora.

Interactions with animals further reinforce the invasive success of *L. camara*. Frugivorous birds such as P. goiavier and A. panayensis, together with fruit bats like C. brachyotis, disperse viable seeds across the landscape, accelerating invasion into new habitats (Buckley et al., 2006). Its flowers attract pollinators, including the Asian honey bee (*Apis cerana*), stingless bees (Tetragonula spp.), and butterflies such as swallowtails (Papilio spp.) and milkweed butterflies (Danaus spp.), providing nectar but potentially disrupting native pollination networks by diverting pollinator visits from indigenous plants (Totland et al., 2006). Most herbivores and livestock, including goats and cattle, avoid *L. camara* due to its toxicity, reducing the grazing value of invaded lands, while dense thickets provide only short-term shelter for small birds (e.g., munias, Lonchura spp.) and invertebrates, contributing to food-web shifts that favor generalist species over specialists.

Despite its largely negative ecological impacts, *L. camara* provides limited ecosystem services in disturbed landscapes. Its flowers serve as a nectar source for pollinators such as Apis cerana and Tetragonula spp., and its fruits provide food for birds and bats, including Pycnonotus spp. and Cynopterus spp., particularly in degraded areas with few alternative resources (Totland et al., 2006). The plant can also contribute to short-term soil stabilization on eroded slopes and roadsides, and it has localized human uses for hedging, traditional medicine, and fuelwood (Day et al., 2003). However, these benefits are generally temporary and context-dependent.

At the ecosystem scale, sustained invasion by *L. camara* results in reduced native biodiversity, with declines in native plants such as Shorea species and associated specialist fauna (Sharma et al., 2005). Landscapes become increasingly homogenized as lantana thickets replace diverse vegetation, and in some environments, increased fuel continuity alters fire regimes, potentially raising fire frequency or intensity. These impacts create positive feedback loops in which dense lantana growth enhances fruit production and animal-mediated seed dispersal, further reinforcing its dominance and spread (GISD, 2023).

Effective management of *L. camara* in the Philippines requires early detection and rapid removal to prevent establishment and spread, using mechanical cutting, uprooting, and targeted herbicide application (Day et al., 2003). Post-removal restoration through replanting native species such as *S.*

contorta, S. polysperma, and *M. tanarius* is essential to suppress reinvasion from the persistent seed bank. Integrated control approaches that combine mechanical, chemical, biological, and community-based strategies, together with long-term monitoring and research, are necessary to evaluate control effectiveness and reduce long-term impacts on native biodiversity (Taylor et al., 2012).

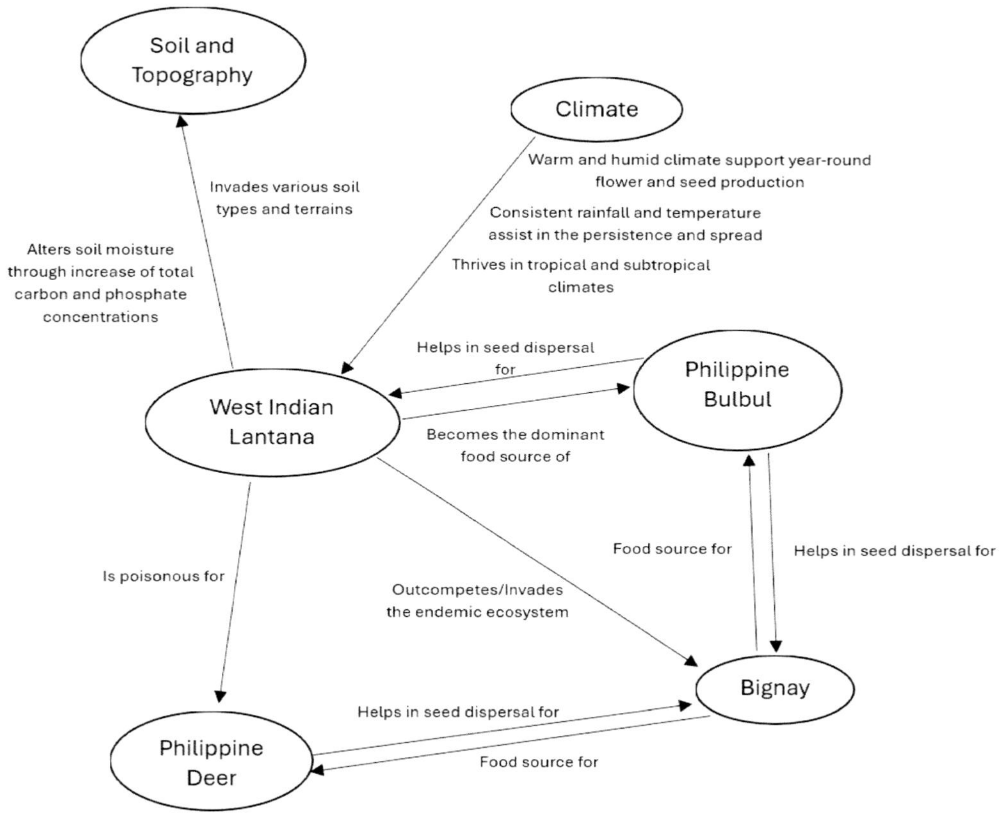

Figure 3. Interactions of *Lantana camara* with various ecosystem components (soil and topography, climate, and native Biota)

The figure above showcases the different interactions between the *L. camara* and various ecosystem components, such as soil and topography, climate, and an example of interactions between native biota. For soil and topography, Kato-Naguchi & Kato's (2025) recent study on the compounds involved in the invasive characteristics of Lantana stated that the presence of the invasive species has altered soil moisture and has actually increased the concentrations of total carbon and phosphate in the soil. Adding to this, resilience and its high reproductive capacity can be cited as one of the reasons why it is able to invade various soil types and terrains, as its toxic and allelopathic compounds reduce herbivory and competition, which in turn, enhances establishment success.

Climate is also considered a critical factor in the spread of the invasive species, whereas climates with distinct wet seasons allow the species to peak in terms of flowering and seed production after substantial rainfall. However, in broadly warm-humid zones, it is shown that they can flower and fruit nearly year-round as the soil moisture rarely becomes limiting (Gardenia.net, 2024).

As for native biota, the inclusion of native species, both of flora and fauna are considered, whereas the Bignay (*Antidesma bunius*), Philippine Bulbul (*Hypsipetes philippinus*,), and Philippine Deer (*Rusa marianna*) are taken as examples. The introduction of the alien species has put a strain on endemic species such as the Bignay, as the present allelochemicals released by the *L. camara* may suppress the regeneration process of indigenous plant species, suppressing their regeneration process and decreasing germination and seedling growth, leading to their decline (Kato-Noguchi & Kurniadie, 2021). As a frugivorous bird, the Philippine Bulbul, as stated in Schabacker and Curio (2022), benefits fruiting trees through the means of seed dispersal via ingestion of seeds as they eat the fruits as a whole. Despite having pentacyclic triterpenoids known as lantadenes (a toxin known to cause poisoning, liver damage, and photosensitivity in herbivores such as those applied in the Philippines, directed at the Philippine Deer), the Philippine Bulbuls, much like other species of birds, have a resistance to plant toxins and unwittingly spread their seeds.

References

Buckley, Y. M., Anderson, S., Catterall, C. P., Corlett, R. T., Engel, T., Gosper, C. R., Nathan, R., Richardson, D. M., Setter, M., Spiegel, O., Vivian-Smith, G., Voigt, F. A., Weir, J. E. S., & Westcott, D. A. (2006). Management of plant invasions mediated by frugivore interactions. *Journal of Applied Ecology*, 43(5), 848–857. https://doi.org/10.1111/j.1365-2664.2006.01210.x

Day, M. D., Broughton, S., & Hannan-Jones, M. A. (2003). Current distribution and status of *Lantana camara* and its biological control agents in Australia, with recommendations for further biocontrol introductions into other countries. *CABI Reviews*, 2003. https://doi.org/10.1079/cabireviews20033155305

Day, M. D., Wiley, C. J., Playford, J., Zalucki, M. P., & Australian Centre for International Agricultural Research. (2003). Lantana: Current management status and future prospects. Australian Centre for International Agricultural Research. http://www.aciar.gov.au

Fern, K. (2025, November 17). *Lantana camara*. Tropical Plants Database. https://tropical.theferns.info/viewtropical.php?id=Lantana+camara

Gardenia.net. (2024, November 30). *Lantana camara* (Bigleaf Lantana). Gardenia. https://gardenia.net/plant-varieties/lantana-camara-bigleaf-lantana

Global Invasive Species Database (2025) Species profile: *Lantana camara*. http://www.iucngisd.org/gisd/speciesname/Lantana+camara

Global Invasive Species Database (GISD). (2023). *Lantana camara*. Invasive Species Specialist Group (ISSG), IUCN. https://www.iucngisd.org

iNaturalist. (2016, September 26). *Lantana camara* (BIS-Fogo). https://www.inaturalist.org/taxa/105418-Lantana-camara

Invasoras.pt. (n.d.). Characterisitics of *Lantana camara*. https://invasoras.pt/en/invasive-plant/lantana-camara-0

Kalaichelvi, K (2020). Eradication of *Lantana camara* and strategies for the control of Lantana in hilly zones. *Indian Journal of Plant Sciences*, 57–60, 57–60. https://tinyurl.com/2mmpev5h

Kato-Noguchi, H., & Kato, M. (2025). Compounds involved in the invasive characteristics of *Lantana camara*. Molecules, 30(2), 411. https://doi.org/10.3390/molecules30020411

Kato-Noguchi, H., & Kurniadie. D. (2021). Allelopathy of *Lantana camara* as an invasive plant. *Plants*, 10(5), 1028. https://doi.org/10.3390/plants10051028

Mishra, P., et al. (2023). Morphology and invasive characteristics of *Lantana camara*. IJNRD, 8(3).

Mridula, Rana, M., Sharma, G., Anshumali, Rana, S., & Sharma, S. (2025). *Lantana camara*: A review on ecology, invasion and management strategies. *International Journal of Research in Agronomy*, 8(2), 141–147. https://doi.org/10.33545/2618060x.2025.v8.i2c.2543

Schabacker, J., & Curio, E. (2022). Notes on Fruit Consumption of the Philippine Bulbul (*Hypsipetes philippinus*) and its quality as a seed disperser. *Silliman Journal*, 43(1). Retrieved from https://sillimanjournal.su.edu.ph/index.php/sj/article/view/242

Sharma, G. P., Raghubanshi, A. S., & Singh, J. S. (2005). Lantana invasion: An overview. *Weed Biology and Management*, 5(4), 157–165. https://doi.org/10.1111/j.1445-6664.2005.00176.x

Taylor, S., Kumar, L., Reid, N., & Kriticos, D. J. (2012). Climate change and the potential distribution of an invasive shrub, *Lantana camara* L. *PLoS ONE*, 7(4), e35565. https://doi.org/10.1371/journal.pone.0035565

Totland, O., Nyeko, P., Bjerknes, A.-L., Hegland, S. J., & Nielsen, A. (2006). The effect of an exotic plant (*Lantana camara*) on pollination networks in Uganda. *Biological Conservation*, 129(3), 404–414. https://doi.org/10.1016/j.biocon.2005.11.003

Mimosa invisa: An Invasive Dynamics and a Complex Species

John Lloyd B. Abejo, Alvin Q. Andrade & Mark June G. Rodil

Species, Details, and Features of the *Mimosa invisa* Species

Name of an Invasive Species

The International Union for Conservation of Nature (IUCN) defines Invasive alien species (IAS) as animals, plants, or other organisms introduced, intentionally or accidentally, into areas outside their natural range, and that negatively impact native biodiversity, ecosystem services, or human economies and well-being. One of the most significant risks to biodiversity is the introduction and/or spread of invasive alien species, which have affected native biodiversity in nearly every ecosystem on Earth (CBD, 2006). Invasive species are highly adaptable and can alter and even disrupt ecosystem functions in newly colonized ecosystems. Invasions by alien plants alter ecosystem services and threaten ecosystem integrity and functioning (Walsh et al., 2016).

Mimosa invisa is a species of particular concern in the Philippines that demonstrates these invasive traits. M. diplotricha C. Wright, often referred to as the enormous sensitive plant, the creeping sensitive plant, or nila grass, is recognized as a synonym of *M. invisa*. In the Philippines, it is called Makahiyang lalaki. Taxonomists classify it under the order Fabales, Class Magnoliosida, and Division Magnoliophyta (Sankaran, n.d.).

The *M. invisa* is a fast-growing, upright shrub and a scrambling climber that can quickly produce dense thickets. Despite its perennial appearance, it is an annual. Bright green, fluffy, alternating leaves with roughly 20 pairs of tiny leaflets each are bipinnate, sessile, opposite, lancelate, acute, 6-12 mm long and 1.5 mm wide, and sensitive to disturbance. The stem can reach a height of three meters, has four angles, is woody at the decumbent base, and has recurved thorns that range in length from three to six millimeters. The corolla is gamopetalous, with twice as many stamens as petals; the inflorescence is a clustered, fluffy, pale pink ball that is about 12 mm wide and appears on short stalks (1 cm tall) in leaf joints. It blooms from August to February, though this may vary by location; in some tropical countries, it blooms year-round. The pods are flat, linear, clustered, 10-35 mm long, 6 mm broad, and covered in tiny prickles. At the ends, they divide transversely into pieces with a single seed. The seeds are shiny, light brown, flat, and ovate, measuring 2-2.5 mm in length and 0.6-1.4 mm in thickness. Seed yield ranges from 8,000 to 212,000 per meter. One thousand seeds weigh about six grams. From September to February, seeds are set. The root has many branches and nodules (Sankaran, n.d).

A detailed description of its morphology further illustrates the species' capacity to spread rapidly and dominate invaded habitats. Native to South America, *M. invisa* is a leguminous woody shrub. It is divided into two subspecies, each further classified into two varieties. *M. invisa* Barneby is the first, and it contains var. invisa, which is indigenous to Paraguay and Brazil. Brazil and Paraguay are also home to macrostachya (Bentham) Barneby. *M. invisa* spiciflora (Karsten) Barneby, which includes a variety, is the second spiciflora, which is indigenous to northern South America—Venezuela's native tovarensis (Bentham) Barneby. In many areas of the United States, *M. invisa* is considered an invasive and noxious plant outside of its natural range. According to Jayasree (2005), there are two types of *M. invisa* based on the presence or absence of spines on the stem and other plant parts. They are *M. invisa* var. invisa, called spiny or thorny mimosa, which is armed with sharp recurved prickles, and *M. invisa* var. inermis, called smooth mimosa, which is unarmed or spineless.

NIAES (2007) described the morphological features of *M. invisa* as having stems that are branches and oblique, with five ridges bearing dense retrorse spines. The bipinnate leaves include 15–30 pairs of leaflets and are somewhat touch-sensitive. The glabrous inflorescence has several blooms that emerge from the axils and the end of the stalk. Each reddish-purple floret has eight stamens. Pods are hispid and spiky. Waterhouse (1994) reported that *M. invisa* is a fast-growing, thorny shrub that is either biennial or perennial, with angular, branching stems that eventually become woody. The leaves are complex, bipinnate, and alternating. A short, thorny stalk arising from a leaf axil bears globular, pink- to purple-colored blooms.

The seed pods are covered with stiff bristles and split at transverse grooves into two to four single-seeded segments. Moreover, *M. invisa* folds its pinnate leaves when touched, but it is less sensitive than some other species, such as *M. pudica*. The leaves fold at nightfall. Unlike in more tropical regions, such as the Philippines, where *M. invisa* flowers year-round, in central and southern Brazil, it flowers only from the end of January to mid-April. From February to the end of May, seeds mature. After that, plants senesce, losing most of their leaves; some green leaves remain at the base of the stem. As stated by Petruzzello (2009), A few M. species are trees, some are woody climbers, and the majority are herbs or undershrubs. They are often prickly. The majority have bipinnate leaves, meaning that the feather-formed leaves themselves have leaflets. Some plants have poisonous roots, while others have skin-irritating compounds. Small, regular flowers characterize the plants and produce legume fruits. In addition to those responsive to physical stimuli, a few species have leaves that are sensitive to light and droop in response to darkness.

The movement of the leaflets within the compound leaves of *M. invisa* is a reaction to external stimuli, including touch, vibration, or thermal changes. This phenomenon is known as seismonasty. A related response triggered by darkness, particularly regarding photoperiodism, is called nyctinasty (Simpson, 2019). It fixes nitrogen and suppresses competing weeds and beneficial plants through its rapid growth (Waterhouse, 1994).

Native Range and Introduction Technique

M. invisa Martius ex Colla is a native of tropical America and is also known as the Giant Sensitive Plant or the Creeping Sensitive Plant. Outside its natural habitat, *M. invisa* was recorded in Java, Indonesia, in 1900 (Soerjani et al., 1987). Jayasree (2005) claims that it has spread to all five inhabited continents in the last century: Asia (Southeast Asia and the Pacific), Australia (Queensland), Africa (Nigeria), and Europe (Cuba). Waterhouse (1994) assessed the global distribution and significance of *M. invisa*. As a result, the plant is widely distributed and plays a highly significant role in the Philippines. Although the plant is not very common in Myanmar, Thailand, Laos, Vietnam, Malaysia, Singapore, and Indonesia, its influence on local vegetation is significant.

According to Sankaran (n.d.), the species is found in South and Southeast Asia, the Pacific Islands, northern Australia, South and Central America, the Hawaiian Islands, portions of Africa, especially Nigeria, and France. It is currently common in Kerala and several northeastern states of India, especially Assam, but its presence in other states remains unknown and warrants further investigation. Specifically, authorities have not yet officially designated *M. invisa* as a weed in the Americas, Western Asia, East Africa, and Europe.

The introduction of *M. invisa* into new areas is often accidental, primarily linked to human activity and the movement of contaminated materials. Seeds can spread through infected soil or crop seeds, as well as through attachment to animal fur, clothing, cars, machinery, and agricultural tools. Remarkably, seeds can remain dormant in the soil for up to 50 years, and dry heat promotes seed germination.

Economic and Ecological Effects of *Mimosa invisa*

M. invisa is an invasive species that is native to tropical America but has gone beyond its natural habitat. Although adapted to its original habitats, that plant has colonized grasslands, fields, plantations, and forest margins, forming dense, often monocultural stands. *M. invisa* has spread quickly, threatening native ecosystems and biodiversity in places like the Philippines. As the infestation spread, *M. invisa* significantly influenced the biodiversity of native plants (Jayasree et al., 2006). Data on the number of active vegetative shoots per unit area, recorded over three consecutive years (2002, 2003, and 2004), showed that *M. invisa* infestation in an area can considerably reduce the population of other native flora. Because of its sharp, recurved thorns, cattle find it difficult to get through the stands and are unwilling to graze on crops infected with it. Because the thorns pierce and cut the workers' hands, harvesting *M. invisa* is difficult. It is prevalent in damp waste areas and along roadways. Crop lands, pastures, sugarcane and pineapple fields, coconut, tea, and rubber plantations were severely impacted. It is a significant weed in Southeast Asia and the Pacific, but it is not a problem in the Americas, Western Asia, East Africa, or Europe (Waterhouse & Norris, 1987). At all contaminated sites, Mimosa's dense growth inhibits the growth, reproduction, and regeneration of native species. All parts of the plant are poisonous when consumed by herbivores. It produces a toxin called mimosin, a non-protein amino acid that can harm the vascular endothelium, cause liver and heart necrosis, and cause anemia in cattle (Sankaran, n.d.).

According to a fact sheet released by the Asia-Pacific Forest Invasive Species Network (APFISN), infestations of *M. invisa* result in higher cultivation expenses, lower crop yields, and decreased land values. It also highlights how difficult it is to eliminate once established, as the plant can withstand a variety of soil conditions. In the Philippines' agricultural areas, it causes significant economic issues. Farmers find it challenging to control contaminated fields, particularly during harvest, due to the fast growth and prickly stems. On a broader economic scale, it contributes to soil degradation, reduced land productivity, and declining land values, particularly in provinces where agriculture is the primary driver of the economy. *M. invisa* disrupts watershed functions by altering soil structure and increasing erosion, which in turn affects water quality and irrigation systems. The combined effects of these disruptions may result in reduced food production, increased imports, and a decline in the competitiveness of Philippine agricultural exports. Additionally, government agencies must allocate funds for invasive species management, thereby diverting resources from other development objectives.

Control and Eradication Measures

M. invisa alters an area's natural biodiversity by growing rapidly, forming dense stands, and strongly competing with native species. If left unchecked, it can encroach on grasslands, woodlands, riverbanks, and farmlands, degrading wildlife habitat quality and restricting the establishment of native flora. Its aggressive growth can disrupt ecological balance and displace native species. Studies from Australia and Southeast Asia report that mechanical removal alone is insufficient, as *M. invisa* vigorously resprouts, underscoring the need for integrated weed management. The effective management requires an approach of different control strategies, including the following;

Mechanical Control

One of the oldest and easiest methods for controlling these species is still mechanical management, particularly on small-scale agriculture or recently infested areas. Young plants are frequently uprooted by hand because it is easier to remove the entire root system when the soil is moist. Only when the roots are completely removed to stop regrowth is this technique the most successful. Slashing and cutting have been commonly used in larger infestations to reduce seed production and lower aboveground plant matter. Still, because of the plant's strong resprouting ability, these methods must be repeated frequently. According to the APFISN (2018), slashing is not advisable, as the weed can easily regenerate from cut stumps, and grazing by animals helps prevent the dominance of *M. invisa*.

Various studies summarized by Jayasree (2005) indicate that frequent soil cultivation significantly reduces the number of viable seeds, with cultivated soil leading to faster seed mortality than undisturbed soil (Roberts, 1962; Roberts & Dawkins, 1967; Roberts & Feast, 1973, as cited by Jayasree, 2005). Mechanical methods such as manual weeding, cutting at ground level, or using tools to uproot plants are commonly practiced. However, the mechanical removal of spiny *M. invisa* is labor-intensive and often results in resprouting if not done properly (Sankaran, 2001; Schatz, 2001; Alabi et al., 2004, as cited by Jayasree, 2005). Timely removal, such as within five weeks after planting, has been shown to support crop recovery and maximize yield.

Biological Control

Biological control, by contrast, can suppress *M. invisa* for extended periods in a sustainable manner. Research has identified numerous natural enemies that can restrict or delay the species' growth. The psyllid Heteropsylla spinulosa is one of the most prominent; it feeds on the sap of the plant, which eventually stunts its growth and decreases its capacity for reproduction. Specific fungal pathogens, such as Phloeospora mimosae and Cercospora species, also contribute to natural regulation by weakening plants through leaf spot diseases.

In the Philippine setting, a fungus isolated from diseased *M. invisa*, Fusarium pallidoroseum, provided excellent control of Mimosa seedlings when sprayed with crude or cell-free culture filtrate. Researchers need to conduct further studies before recommending the use of the fungus to control *M. invisa*. These control strategies are essential and valuable in large or remote areas where chemical and mechanical interventions are feasible.

Numerous insect species have been identified as effective biocontrol agents for *M. invisa*. The psyllid haplotypes without impacting neighboring plants, spinulosa decreases plant growth, tip elongation, and seed production (Ablin, 1992, 1993a, 1993b; Waterhouse, 1994, as cited by Jayasree, 2005). Western Samoa, Pohnpei, and Papua New Guinea have all successfully released it. The moth Psygida walkeri exhibits host specificity and inhibits flowering and seed production by feeding on leaves, buds, pods, and stems—Scamurius sp., the coreid insect. Feeds on shoots and collapses developing tips to prevent vegetative growth and flowering (Garcia, 1984; Wild, 1986, 1987, as cited by Jayasree, 2005).

Fungal pathogens have also been widely explored for their biocontrol potential. C. cassiicola, a stem spot fungus common in Queensland, Papua New Guinea, and Western Samoa, can severely damage *M. invisa*, causing leaflet shedding and stem dieback as lesions spread (Haseler, 1984; Wilson & Ablin, 1991; Ablin, 1992, as cited by Jayasree, 2005). From Venezuela, *Uredo mimosae-invisae* attacks *M. invisa* (Evans, 1999, as cited by Jayasree, 2005). In the Philippines, F. pallidoroseum effectively controlled seedlings in laboratory and field trials, though potential health risks must be studied before widespread use (Mabbayad & Watson, 2000, as cited in Jayasree, 2005). C. canescens is also considered a potential bioagent, pending detailed host specificity studies (Waterhouse, 1994), as cited by Jayasree (2005).

Chemical Control

When applied according to the recommended rates, various herbicides such as glyphosate (0.75 kg/ha), paraquat (0.5 kg/ha), diuron (2–4 kg/ha), a combination of acetochlor and atrazine (0.92 + 0.63 kg/ha), starane (1.3–1.5 L/ha), a mixture of atrazine and metolachlor (0.82 + 1.68 kg/ha), and atrazine at 500 g a.i. (4–6 L/ha) can be effective. It is best to time the applications before the onset of flowering and fruiting for optimal results. However, the effectiveness of these herbicides is often short-lived, necessitating reapplications as weeds regrow. The actual performance of chemical control can vary significantly across locations, despite general recommendations.

Chemical control of *M. invisa* has produced mixed results. Applications of 2,4-D and dinoseb were largely ineffective in Brazil (Lew, 1993, as cited by Jayasree, 2005), and chemical or mechanical methods are often expensive and provide only short-term effects (Muniappan & Virakthamath, 1993, as cited by Jayasree, 2005). Certain herbicides, including alachlor, atrazine, and oxyfluorfen, have been effective against smooth *M. invisa* in Thailand (Pornprom et al., 2002, as cited in Jayasree, 2005). Trials at Kerala Agricultural University found that glyphosate was the most effective, achieving up to 50% control at 0.2 kg/ha and excellent suppression at 0.4 kg/ha (KAU, 2003, as cited by Jayasree, 2005).

Several studies support the need for integrated weed management involving all three control strategies. Research from Papua New Guinea and Australia indicates that mechanical control alone is often inadequate due to rapid resprouting. In contrast, chemical treatments without follow-up measures tend to yield only temporary suppression. Biological control agents, such as H. spinulosa, have demonstrated promising efficacy in limiting plant growth, particularly in areas where herbicide use is restricted or costly (Auld, B. A., et al., 1996).

Original Field Photos

Fig. 1 *M. invisa* leaflets, flowerhead, and stem

Fig. 2 Dense, tangled growth of *M. invisa* showing its climbing and sprawling habit

Figure 2 shows the rapid growth of *M. invisa*, highlighting its tendency to climb and its sprawling growth habit. The plant's dense cover of sharp, recurved thorns makes removal difficult and hazardous.

Fig. 3 Vacant lot heavily infested by dense, sprawling *M. invisa*.

Figure 3 illustrates the environment in which *M. invisa* thrives, particularly in open areas such as road-sides, where it aggressively spreads and colonizes disturbed habitats. This plant poses a threat to humans and animals, as it can obstruct roads and pathways and cause injuries from its sharp, thorny stems.

Concept Map

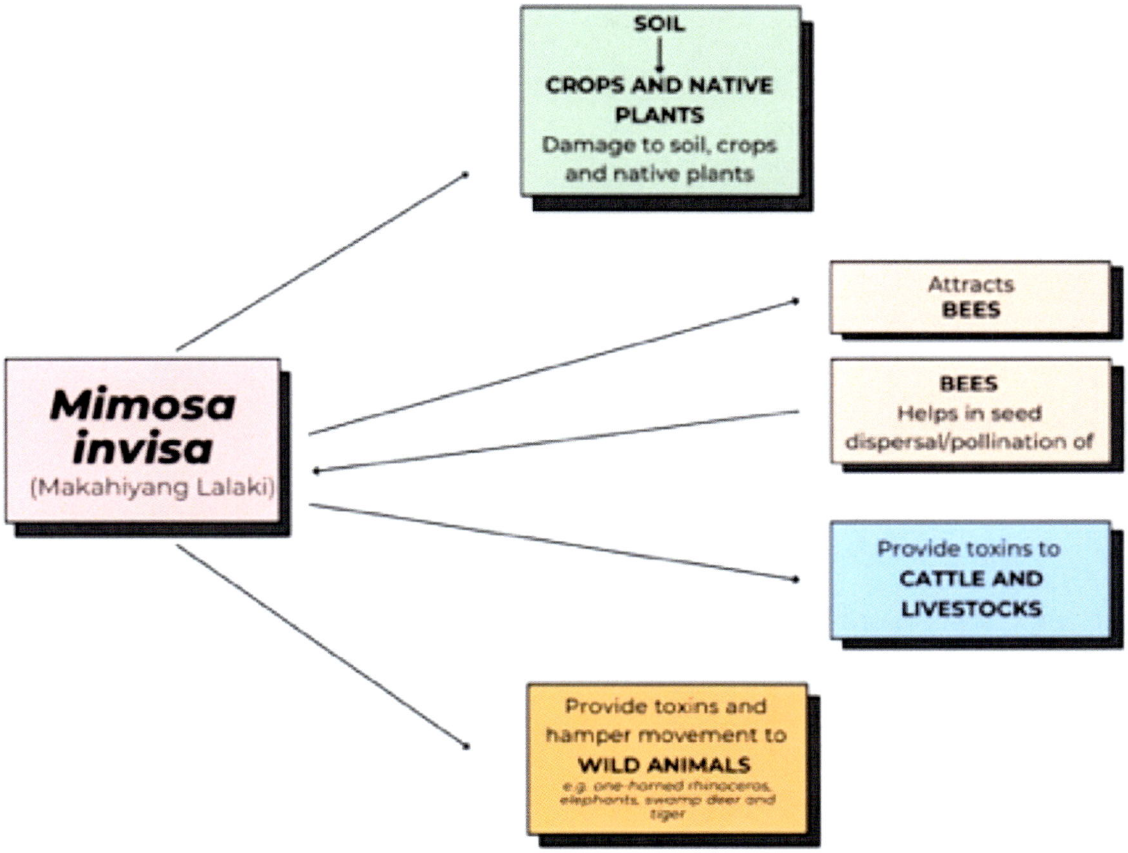

M. invisa produces flowers that attract bees, highlighting its role in the local pollination network. Bees assist in seed dispersal, and the seeds remain dormant for up to 50 years. *M. invisa* is a significant threat to forest ecosystems, agricultural land, and pastures. It causes heavy damage to crops such as sugarcane, coconut, rubber, cassava, tea, pineapple, and upland rice. Thick Mimosa growth prevents the regeneration, reproduction, and growth of indigenous species across all infested areas. All parts of the plant are toxic to herbivores if ingested. It produces a toxin (mimosin, a non-protein amino acid) that can cause vascular endothelial damage, heart and liver necrosis, and anemia in cattle.

The tangled and thorny growth of Mimosa hampers movement and access to food and other resources for wild animals like the one-horned rhinoceros (*Rhinoceros unicornis*), an endangered species, Asian elephant (*Elephas maximus*), swamp deer (*Cervus duvauceli*), and tiger (*Panthera tigris*) in Kaziranga National Park in north-east India. In Australia, the weed chokes out cane and other crops and grassland, causing crop and pasture loss. Crops infested with mimosa are difficult to harvest due to the thorns. Increased cultivation costs, reduced crop yields, crop losses, reduced land value, and soil degradation are the main economic impacts of Mimosa (APFISN, 2018) .

References

Asia-Pacific Forest Invasive Species Network. (2018). *Mimosa diplotricha* Fact Sheet. Retrieved from https://apfisn.org

Auld, B. A., & Martin, P. M. (1996). The autecology of *Mimosa invisa Mart. Ex Colla* in the humid tropical lowlands of Papua New Guinea. *Australian Journal of Agricultural Research, 26*(4), 689–702.

Convention on Biological Diversity (2009). What are Invasive Alien Species? https://www.cbd.int/idb/2009/about/what

IUCN. (n.d.). Invasive Alien Species. https://iucn.org/our-work/topic/invasive-alien-species

Jayasree, P.K. (2005). Biology and management of *Mimosa invisa Mart*. in Kerala [Unpublished doctoral dissertation]. Kerala Agricultural University.

Jayasree, P. K., Abraham, C. T., & George Thomas, C. (2006). Adverse effects of *Mimosa invisa Mart*. infestation on the floral biodiversity. *Indian Journal of Weed Science*, 38(1 & 2).

National Institute for Agro-Environmental Sciences, 2007. Asian-Pacific Alien Species Database. https://www.naro.affrc.go.jp/archive/niaes/techdoc/apasd/Mimosa

Petruzzello, M. (2009). Mimosa. Britannica. https://www.britannica.com

Sanakaran, K.V., n.d. Mimosa Diplotricha: Giant Sensitive Plant, Asia-Pacific Forest Invasive Species Network. https://www.fao.org/forestry-fao/

Simpson, M. G. (2019). Plant Systematics. Science Direct, https://www.sciencedirect.com/topics/agricultural-and-biological-sciences/mimosa

Soerjani, M., Kostermans, A.J.G.H., and Tjitrosoepomo, G. (1987). Weeds of Rice in Indonesia. https://www.cabidigitallibrary.org/doi/full/10.5555/19901181858

Waterhouse, D.F. (1994). Biological Control of Weeds: South East Asian Prospects, https://www.aciar.gov.au/sites/default/files/legacy/node/2160/mn26_pdf_6381 9.pdf

Waterhouse, D.F. & Norris, K.R. (1987). Biological control: Pacific Prospects, Inkata.

Walsh, J. R., Carpenter, S. R., & Vander Zanden, M. J. (2016). Invasive species triggers a massive loss of ecosystem services through a trophic cascade. Proceedings of the National Academy of Sciences, 113(15), 4081–4085. https://doi.org/10.1073/pnas.1600366113.

Water Hyacinth (*Pontederia crassipes*): Invasion and Management

Ronalyn L. Mañosca, Mary Anne G. Causapin & Jaeshienne Therese K. Payumo

I. Introduction

Biological invasions represent a major driver of ecological change across terrestrial and aquatic systems. Despite differences in habitat structure, invasive plant species frequently succeed through a shared suite of functional traits that transcend ecosystem boundaries. Water hyacinth (*Pontederia crassipes* Mart.) (formerly *Eichhornia crassipes* [Mart] Solms), is one of the world's most problematic aquatic invaders that exhibits ecological strategies analogous to those observed in invasive plants of forest margins and disturbed terrestrial landscapes. These strategies, such as high relative growth rates, allelopathic interference, and strong affinity for anthropogenically disturbed environments, explain both its rapid expansion and its persistent dominance in invaded freshwater ecosystems.

A central trait underlying the invasiveness of Water hyacinth (*P. crassipes* Mart.) is its exceptionally high biomass accumulation and clonal propagation capacity. The species primarily reproduces vegetatively through stolon, allowing exponential population increase under favorable environmental conditions. Empirical studies have demonstrated that water hyacinth can rapidly increase surface coverage and biomass in nutrient-enriched waters, often outpacing native macrophytes (Gopal, 1987; Wilson et al., 2005). This growth strategy parallels invasive terrestrial plants that exploit canopy gaps and edge environments, where rapid height or lateral expansion enables early monopolization of resources. In aquatic systems, dense hyacinth mats restrict light penetration, suppress submerged vegetation, and reduce dissolved oxygen concentrations, leading to altered trophic dynamics and habitat degradation (Villamagna & Murphy, 2010).

Beyond competitive exclusion through physical dominance, water hyacinth also employs chemical interference mechanisms that enhance its invasive success. Research indicates that the species releases secondary metabolites into surrounding waters that negatively affect the growth and physiological performance of co-occurring aquatic plants and phytoplankton (Malik, 2007). This allelopathic capacity mirrors mechanisms observed in invasive forest species that modify soil chemistry to inhibit native plant establishment. In both aquatic and terrestrial contexts, allelopathy functions as a non-resource-based competitive strategy that restructures community composition and reinforces invasive dominance.

Equally important is water hyacinth's strong association with disturbed and nutrient-enriched environments, a hallmark of invasive species across ecosystems. The plant thrives in eutrophic water bodies receiving high nitrogen and phosphorus inputs from agricultural runoff, sewage discharge, and urban development (Villamagna & Murphy, 2010). These conditions are ecologically analogous to disturbed forest edges, degraded lands, and human-modified landscapes that facilitate terrestrial plant invasions. Water hyacinth's (*P. crassipes* Mart.) physiological tolerance to fluctuating water levels, reduced water quality, and chemical stressors allows it to persist where native species experience reduced fitness, thereby accelerating invasion processes (Wilson et al., 2005).

From a cross-ecosystem perspective, Water hyacinth (*P. crassipes* Mart.) exemplifies how invasive plants exploit resource pulses, disturbance regimes, and biotic resistance gaps irrespective of habitat type. Its success highlights the importance of functional traits rather than ecosystem classification in understanding invasion dynamics. Consequently, management approaches that focus solely on species removal without addressing underlying drivers such as nutrient loading and habitat disturbance are unlikely to achieve long-term control.

In conclusion, Water hyacinth (*P. crassipes* Mart.) demonstrates that invasive species across forest and aquatic boundaries rely on shared ecological strategies. Its rapid clonal growth, allelopathic interactions, and adaptation to disturbed environments reflect generalizable invasion mechanisms observed across ecosystems. Recognizing these shared traits provides a conceptual framework for predicting invasion risk and developing integrated, ecosystem-based management strategies at landscape and watershed scales.

II. Lessons in Management and Awareness

The case of Water hyacinth (*P. crassipes* Mart.) provides important lessons in environmental management and public awareness. One key lesson is the importance of early detection and rapid response. Water hyacinth is known for its extreme high growth rate, and delayed intervention often leads to severe ecological and economic impacts. Effective monitoring systems, including the use of remote sensing combined with GIS, and community-based surveillance, are essential to prevent small infestations from expanding into large-scale invasions (Villamagna & Murphy, 2010).

Another major lesson is the need for integrated and adaptive management strategies. Mechanical, chemical, and biological control methods have all been applied to manage water hyacinth. However, no single approach has proven universally effective. Successful management requires combining multiple control techniques while considering site-specific ecological conditions and socio-economic constraints, as well as continuous monitoring and evaluation of management outcomes (Gopal, 1987; Julien, Hill, & Center, 2001).

The invasion of water hyacinth also illustrates the significant role of human activities in species introduction and spread. Originally introduced as an ornamental plant, its widespread distribution highlights the unintended consequences of poorly regulated species introductions. This underscores the necessity of strict biosecurity measures, risk assessments, and policy enforcement to prevent future invasions (Mack et al., 2000).

From an awareness perspective, water hyacinth demonstrates the importance of public education and stakeholder involvement. Local communities often perceive the plant as harmless or useful, which can reduce support for control initiatives. Increasing awareness of its ecological impacts including biodiversity loss, reduced water quality, and disruption of fisheries can foster community participation and improve the effectiveness of management programs (Shanab et al., 2018).

In the Philippines, Department of Environment and Natural Resources (DENR) leads community control through fisherfolk harvesting water hyacinth for cash incentives in Taytay, Rizal, targeting Laguna Lake and Pasig River to protect Manila Bay, integrating manual and mechanical methods. Policy under National Invasive Species Strategy and Action Plan (2020-2030) prioritizes it as a top noxious weed, with task forces for clearing and monitoring (e.g., PRCMO in Ilugin River). Education involves local awareness on impacts like mosquito breeding and fisheries disruption, reforestation ties indirectly through watershed rehabilitation, though aquatic focus limits direct forest links, and no explicit hyacinth-reforestation programs found. Multi-stakeholder governance with LGUs and river rangers ensures adaptive strategies (Apuli & Española, 2022).

Lastly, the management of Water hyacinth (*P. crassipes* Mart.) highlights that invasive species control is not solely a technical issue but also governance and institutional challenge. Sustainable management requires collaboration among scientists, government agencies, local communities, and policy-makers. This reinforces the broader lesson that invasive species management must integrate scientific evidence with social, economic, and institutional considerations (Simberloff et al., 2013).

Fig. 1 Water hyacinth (P. crassipes Mart.) covering the Lakeshore of Pakil Laguna

Fig. 2 Water hyacinth (P. crassipes *Mart.*) with flowers

Invasive Species Concept Map

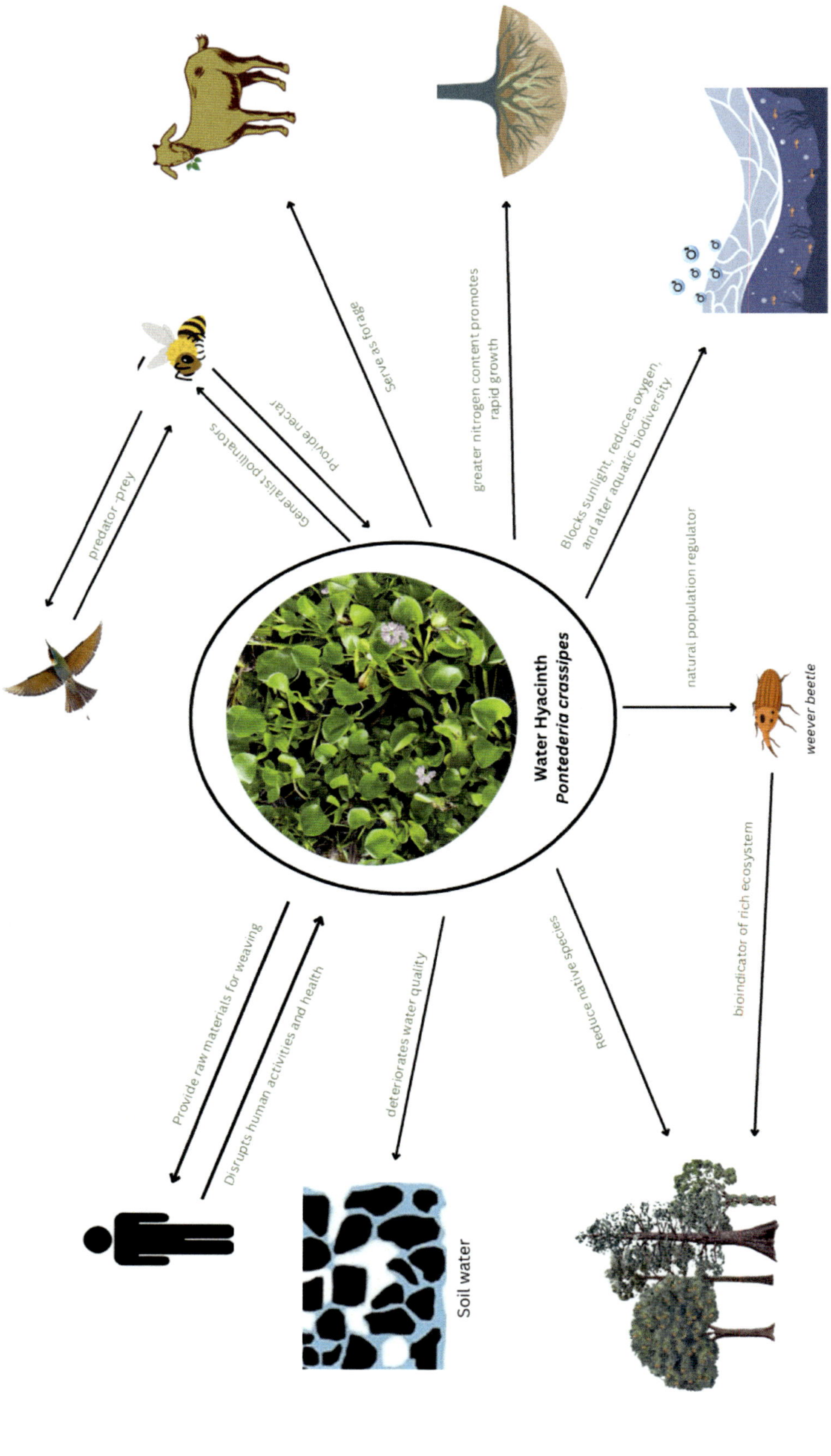

Water Hyacinth Water hyacinth (*P. crassipes*) is a free-floating aquatic plant native to the Amazon Basin that has spread widely across tropical and subtropical freshwater systems such as rivers, lakes, and wetlands (Gopal, 1987; Villamagna & Murphy, 2010). Water hyacinth (*P. crassipes* Mart.) invasions detrimentally affect hydrology, environment, and human well-being by disrupting daily activities and health; increased evapotranspiration can upset local water balances and potentially rainfall patterns, flow reductions promote sedimentation and deoxygenation, weed mats shade lakes reducing light and increasing turbidity, and these changes culminate in lower aquatic populations and higher disease vectors (Harun et al., 2021). These infestations also disrupt transportation, fishing, and irrigation, while providing breeding grounds for mosquitoes and snails that transmit diseases (Dersseh, Sahile, & Enyew, 2019).

Growth and spread are influenced by environmental factors such as temperature, nutrients, and salinity. Warm conditions and high nitrogen and phosphorus promote growth, while higher salinity suppresses it (Patel, 2012; Coetzee, Hill, & Byrne, 2017). Thick mats increase evapotranspiration, reduce oxygen and light in the water, and disturb water flow, negatively affecting fish and other aquatic life (Villamagna & Murphy, 2010). Although pollinated by bees and other insects, Water hyacinth (*P. crassipes* Mart.) reproduces mainly through vegetative growth, allowing rapid expansion (Wilson, Holst, & Rees, 2005). Biological control using weevil beetles (*Neochetina eichhorniae* and *N. bruchi*) can reduce plant biomass but takes time to be effective (Julien, Hill, Center, & Ding, 1999; Center, Hill, Cordo, & Julien, 2002). Chemical control is limited due to environmental risks, while mechanical removal is costly and can create biomass disposal problems (Dersseh et al., 2019) Harvested biomass of Water hyacinth (*P. crassipes* Mart.) can be converted into useful products such as handicrafts, biofuel, and compost, turning waste into economic opportunities to local communities (Jacinto & Lirio, n.d.). Water hyacinth invasions greatly reduce biodiversity by changing habitat structure and resource availability. The thick mats on the water surface block sunlight and oxygen, disrupting the balance of the ecosystem and harming native aquatic plants and animals. This leads to lower species richness, evenness, and overall ecosystem stability.

References

Apuli, A. & Española, A. S. (2022). *A practical proposed use of the growing supply of water hyacinth in Cardona, Rizal recycled into concrete bricks and its application as sidewalk paving materials to address pedestrian safety.* [Unpublished thesis.] https://doi.org/10.13140/RG.2.2.28048.38409/1

Center, T. D., Hill, M. P., Cordo, H., & Julien, M. H. (2002). Water hyacinth. In R. G. Van Driesche. (Eds.), *Biological control of invasive plants in the eastern United States* (pp. 41–64). USDA Forest Service.

Coetzee, J. A., Hill, M. P., & Byrne, M. (2017). *Biological control of water hyacinth in South Africa. African Entomology, 25*(2), 318–326.

Dersseh, W. M., Sahile, S., & Enyew, A. (2019). Impacts of water hyacinth on ecosystem services and human livelihoods in Lake Tana, Ethiopia.

Gopal, B. (1987). Water Hyacinth. Elsevier Science Publishers.

Harun, I., Pushiri, H.., Aiman, A., & Zulkeflee., Z. (2021). Invasive Water Hyacinth: Ecology, Impacts and Prospects for the Rural Economy. Plants (Basel), 10(8), https://doi.org/10.3390/plants10081613

Heliyon, 5(10), e02618. Gopal, B. (1987). Water Hyacinth. Elsevier.

Julien, M. H., Hill, M. P., & Center, T. D. (2001). Biological control of water hyacinth. In Biological and integrated control of water hyacinth, Eichhornia crassipes (pp. 8-20). ACIAR Proceedings.

Mack, R. N., Simberloff, D., Lonsdale, W. M., Evans, H., Clout, M., & Bazzaz, F. A. (2000).

Biotic Invasions: Causes, epidemiology, global consequences, and control. Ecological Applications, 10(3), 689-710. https://doi.org/10.1890/1051-0761(2000)010[0689:BICEGC]2.0.CO;2

Malik, A. (2007). *Environmental challenge vis-à-vis opportunity: The case of water hyacinth. Environment International, 33*(1), 122–138.

Shanab, S. M. M., Shalaby, E. A., Lightfoot, D. A., & El-Shemy, H. A. (2018). *Allelopathic effects of water hyacinth (Eichhornia crassipes). PLoS ONE, 13*(4), e0195226. https://doi.org/10.1371/journal.pone.0195226

Simberloff, D., Martin, J. L., Genovesi, P., Maris, V., Wardle, D. A., Aronson, J., Vilà, M. (2013). *Impacts of biological invasions: What's what and the way forward. Trends in Ecology & Evolution, 28*(1), 58–66. https://doi.org/10.1016/j.tree.2012.07.013

Villamagna, A. M., & Murphy, B. R. (2010). *Ecological and socio-economic impacts of invasive water hyacinth (Eichhornia crassipes). Freshwater Biology, 55*(2), 282–298.

Wilson, J. R. U., Holst, N., & Rees, M. (2005). *Determinants and patterns of population growth in water hyacinth. Aquatic Botany, 81*(1), 51–67.

Williams, A. E., Duthie, H. C., & Hecky, R. E. (2005). *Water hyacinth in Lake Victoria: Why did it vanish so quickly and will it return? Aquatic Botany, 81*(4), 300-314.

Big-Leaf Mahogany (*Swietenia macrophylla*): Spread and Control

Earl Joseph Tinamisan & Roan L. Gendrano

Species History, Information, and Characteristics

1. Invasive Species Name and Description

Scientific Name: *Swietenia macrophylla* King

Family: Meliaceae

Common names: Big-leaf mahogany, large-leaved mahogany, broad-leaf mahogany, Honduras Mahogany

Species Description

Size: A large tree, typically reaching 30-40 m in height, sometimes up to 60 m under very favorable conditions;

Bole (stem): is often branchless for 18-25 m with diameters up to ~1.5-2 m (150-200 cm) in good sites; it develops buttresses up to several meters.

Leaves: Pinnate (paripinnate) compound leaves; typically 2–8 pairs of leaflets (commonly 3–6 pairs), each leaflet measuring about 8-13 (sometimes more) cm × ~3-4(-5.5) cm; young foliage sometimes larger; leaflets are ovate to lanceolate, slightly asymmetrical (one side often more curved).

Flowers: Inflorescences are panicles/supra-axillary, about 10-20 cm long; sepals and petals fringed/hairy in some instances; petals greenish-white to white; moderately fragrant.

Fruit & Seeds: Woody capsules (seed pods), often upright ("sky fruit" in some common parlance) that split via valves (often 5) to release many winged seeds; seed capsules are large, often 10-22 cm length; seeds flat/winged, aiding dispersal by wind.

Bark & Wood: Bark of mature trees is dark grey to greyish-brown, deeply fissured or flaky; inner bark tends toward red-brown/pinkish red. Wood (heartwood) is valued: reddish, fine-to-medium texture, straight-to slightly wavy grain; used for high quality timber (furniture, veneers, musical instruments, etc.).

Site Conditions/ Preferences: Altitudinal range up to ~1500 m, best in warm humid to sub-humid tropics; prefers well-drained soils (medium to heavy), fertile alluvial soils, tolerates some seasonal dryness but needs sufficient moisture; prefers full sun to partial shade, moderate water requirements.

2. Native Range and Method of Introduction

Native range: Tropical America. Specifically, from southern Mexico, through Central America (Belize, Guatemala, Honduras, Nicaragua, Costa Rica, Panama), and in northern South America (Colombia, Venezuela, Ecuador, Peru), plus across parts of the southern Amazon in Bolivia and Brazil.

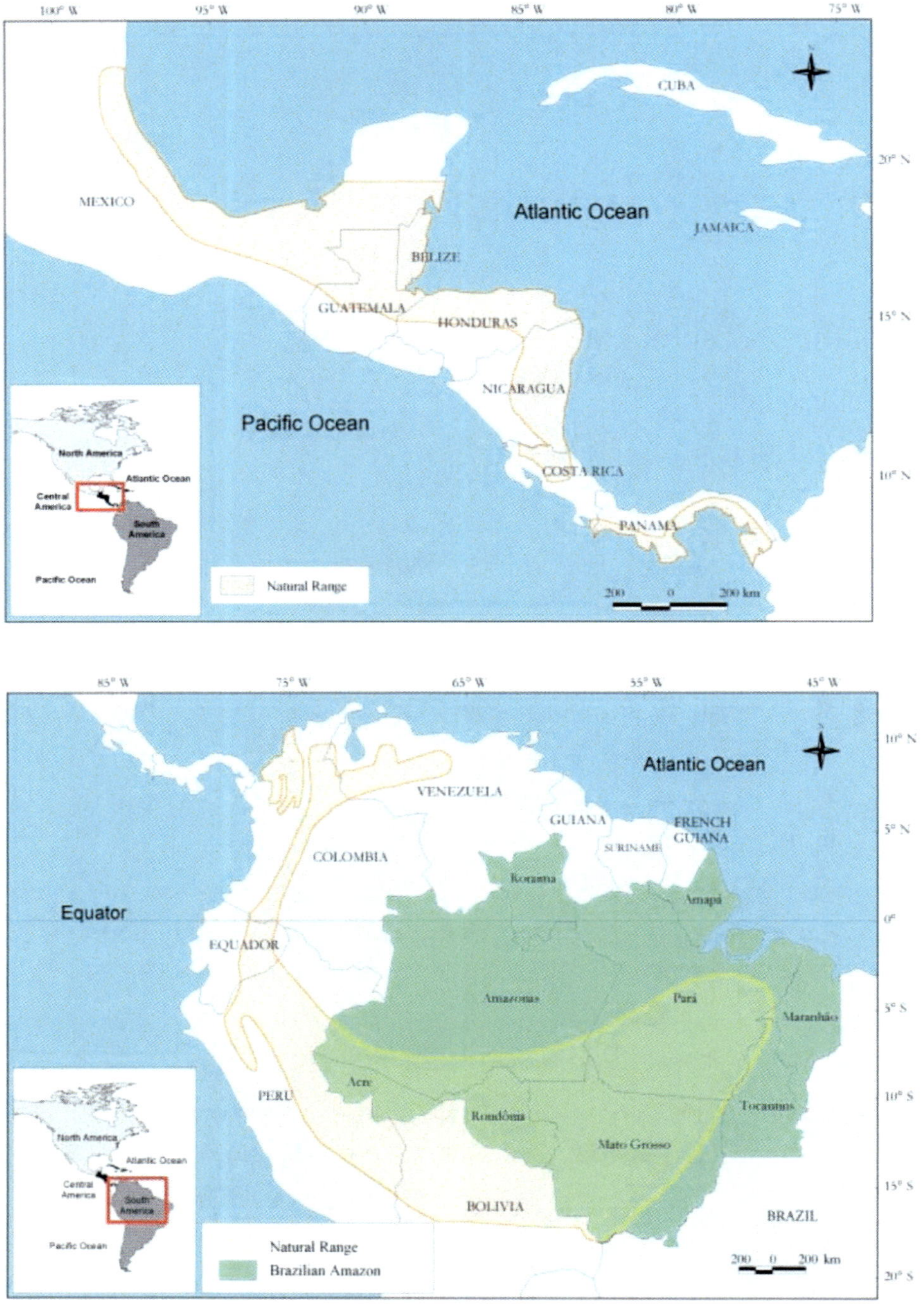

Fig. 1 Native range of *Swietenia macrophylla* (Grogan et al 2002)

Introduced / Naturalized areas: Widely planted outside its native range as a source of valuable timber (mahogany). Naturalized in secondary forest throughout the Philippines as well as in many other tropical countries.

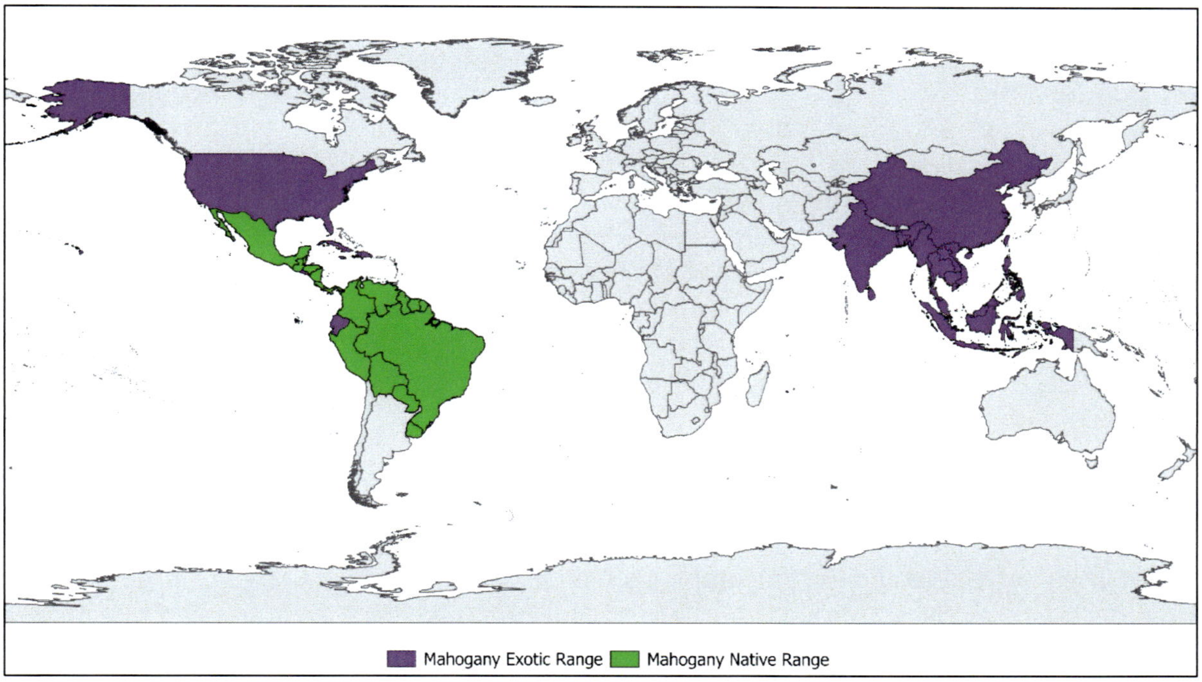

Fig. 2 Distribution map of *Swietenia macrophylla*

Conservation status: Endangered (IUCN) (it is very likely to become extinct in their known native ranges in the near future).

Regulation: CITES Appendix II (meaning international trade is controlled).

3. Ecological and/or Economic Impacts

Ecological Impacts

Big-leaf Mahogany (*Swietenia macrophylla*) exhibits a well-defined reproductive cycle in its native range. In the Amazon Basin, flowering occurs between July and November, spanning the latter part of the dry season and the onset of the rains. Fruit develops throughout the rainy season and reaches full size by mid-season, although seed maturity is achieved only by the beginning of the next dry season, typically between May and August. As the tree shed part or all of its leaves at the start of the dry period, mature fruit capsules open, facilitating wind-driven seed dispersal. Mahogany seeds possess a distinctive thin, tail-like wing that enables autorotation, allowing dispersal distances of up to 500 meters from the parent tree. Mahogany trees bear both male and female flowers, with male flowers occurring at a ratio roughly ten times higher than female flowers. Although the flowers are morphologically similar, functional dioecy has been reported in some individuals. In mixed inflorescences, male flowers open earlier than female flowers, increasing the likelihood of cross-pollination; nevertheless, self-pollination may occur under certain conditions. Flowering and fruiting occur annually, and trees begin to fruit reliably at around 15 years of age (Orwa et al., 2009, Valera et al., 2003).

Across its native range in tropical America, Big-leaf Mahogany (*S. macrophylla*) occupies a broad ecological niche, occurring from pine savannah margins to mature rainforests. Although capable of growth in varied forest types, it attains greatest abundance in mixed hardwood forests, especially along riverbanks and on fertile alluvial soils. Natural populations typically occur at low densities, often fewer than 4–8 trees per hectare, and are usually dispersed or found in small clusters. The species is known to function as a pioneer in degraded agricultural landscapes and is capable of outcompeting other plantation species; for instance, it has been shown to surpass teak in mixed stands. In the Philippine context, the species has demonstrated notable wind firmness and resistance to cyclone damage, making it an attractive plantation species. Despite its silvicultural advantages, Big-leaf Mahogany (*S. macrophylla* is now regarded as one of the most ecologically threatening invasive tree species in the Philippines. Coracero (2023) documented its spread in the buffer zones of the Mt. Banahaw-de-Nagcarlan Protected Landscape on Luzon Island, where 1,591 individuals were recorded in mixed land-use areas and along roadsides. These trees were originally introduced in 1991 by the Department of Environment and Natural Resources for reforestation. Although no individuals were found inside the protected core zone, the study highlighted the absence of a coordinated management strategy and emphasized the urgency of multisectoral planning to prevent further encroachment into native forests.

The invasiveness of Big-leaf Mahogany (*S. macrophylla*) in the Philippines is attributed to its capacity for long-distance wind dispersal, rapid growth, and ability to dominate disturbed forest areas, which collectively impede regeneration of native species. Management recommendations include mechanical removal of seedlings and saplings before reproductive maturity, targeted chemical treatment of cut stumps to prevent regrowth, and potential exploration of biological control approaches. However, Joshi (2006) notes that biological control specific to Big-leaf Mahogany (*S. macrophylla*) is not available, though theoretical models from other invasive plant management systems offer conceptual frameworks for reducing competitive dominance. Given these constraints, early detection, sustained physical removal, and integrated strategies are viewed as essential components of invasive species mitigation.

Table 1 How *S. macrophylla* Affects Other Biological Organisms / Ecosystem Economic Impacts

Impact Category	Description
Competition with native tree species	*S. macrophylla* outcompetes native pioneer and climax species (e.g., dipterocarps) for light, soil nutrients, and space. Fast germination and abundant seeds allow it to dominate regenerating forests and crowd out native trees. Regular annual seed production gives it an advantage over species with irregular seeding cycles.
Reduced native regeneration / biodiversity	Fewer native seedlings occur under or near mahogany stands due to suppression or shading. Understory vegetation becomes sparse ("depauperate"), leading to reduced species richness and overall biodiversity.

Allelopathy / chemical effects	Reports suggest *S. macrophylla* may release allelopathic chemicals (from leaves or litter) that inhibit germination or growth of surrounding plants. Its leaf litter and decomposition can alter soil nutrients in ways unfavorable to native species.
Alteration of habitats for fauna	Declines in native plant species reduce food, shelter, and foraging resources for wildlife. Mahogany-dominated forests offer less variety in fruits, flowers, and foliage, lowering habitat heterogeneity. Such stands are sometimes described as "green deserts."
Rapid colonization in disturbed/mixed land-use areas	Mahogany can spread beyond planting sites, proliferating along roadsides, mixed land-use zones, and edges of protected areas.
Potential effects on soil moisture / hydrology	Dense stands of large, deep-rooted mahogany trees may alter soil moisture balance, water uptake, shading, and evapotranspiration, affecting conditions for ground-layer plants (though documentation is limited).

Big-leaf Mahogany (*S. macrophylla*) is the most economically valuable timber species harvested in the Brazilian Amazon, often fetching prices up to four times higher than those of other timber species. First-quality sawn mahogany can be exported for around US$1,800 per cubic metre from major ports like Belém and Paranaguá to North American and European markets. A mature mahogany tree with an 80 cm diameter can yield over 2 cubic metres of sawn wood, producing more than US$4,100 in international value when the timber is of high quality. Despite these high export prices, the profits from mahogany logging are distributed very unevenly throughout the supply chain. Smallholder farmers and Indigenous peoples, who often have legal or customary rights to the trees, usually receive just US$6–24 per tree—if they are paid at all. Laborers on the ground, such as woodsmen, chainsaw operators, and logging teams, also receive low pay despite doing the most dangerous and physically demanding work. Most of the financial gains go to intermediaries who manage the funding, extraction, processing, and export of mahogany (Veríssimo et al., 1995).

Big-leaf Mahogany (*S. macrophylla*) high market value stems from a combination of aesthetic and physical properties. Its attractive coloration and grain pattern, combined with its notable strength-to-weight ratio, natural resistance to decay, and ease of machining, make it a preferred material for luxury furniture, interior paneling, high-end musical instruments, and yacht construction. These attributes have incentivized loggers to expand deeper into remote Amazonian forests, often reaching areas previously untouched by mechanized logging. In many cases, logging crews rely on small aircraft to locate mahogany crowns from above, targeting isolated stands far from existing infrastructure. Without proper market knowledge and negotiation capacity, landowners and local communities are vulnerable to receiving far below the fair value of the resource (Grogan, 2008).

Big-leaf Mahogany (*S. macrophylla*) has long held a prominent position in Philippine forestry. Although not native to the archipelago, its relatively fast growth, excellent wood qualities, and adaptability have made it a favored species for plantation forestry, reforestation, and agroforestry systems. Big-leaf Mahogany (*S. macrophylla*) is economically important in the local timber industry. Its wood is prized for its strength, workability, and aesthetic properties, making it suitable for furniture, cabinet-making, and other high-quality wood products. In northeastern Mindanao, commercial lumbers of Big-leaf Mahogany (*S. macrophylla*) were tested for mechanical properties, showing competitive bending strength (modulus of rupture) and elasticity (modulus of elasticity) compared to other timber species (Cossid et al 2025).

Because of its commercial value, Big-leaf Mahogany (*S. macrophylla*) plantations offer an income source for rural landowners. Plantation-grown Big-leaf Mahogany (*S. macrophylla*) can be harvested for sawlog production, lumber, or value-added products. In institutional settings, such as university campuses, mahogany forests have even been proposed as a basis for carbon-mitigation plans: one study modeled carbon storage in a mahogany stand and argued that harvesting for furniture (rather than charcoal) yields both economic benefits and lower carbon emissions (Creencia et al 2021).

4. Control and Eradication Measures

Mechanical Control

Mechanical control such as manually uprooting seedlings and saplings or cutting mature trees, often forms the first line of intervention as mahogany reproduces prolifically and can establish dense populations when left unmanaged (Coracero, 2023). According to Baguinon et al (2005) mechanical methods alone are labor intensive and may be insufficient if follow up monitoring is lacking, as mother trees can produce hundreds of seeds annually.

Biological Control

Biological control for Big-leaf Mahogany (S. macrophylla) in the Philippines is very limited and no formal biocontrol agents in use, though the natural herbivory observed in other regions provides significant potential to find specialized insects or pathogens which to prevent ancillary effects would require intensive ecological evaluations (Norghauer et al, 2006).

Chemical Control

Chemical control is impregnation of cut stamps or basal bark with systemic herbicides should supplement mechanical removal by preventing sprouting but must be carefully considered in order to prevent damage to surrounding native vegetation especially in sensitive habitats such as protected areas (Genovesi & Monaco, 2013).

Integrated Control and Eradication Management

Big-leaf Mahogany (S. macrophylla) control and eradication measures require an integrated approach combining the biological, mechanical and chemical strategies to reduce its spread and ecological impact effectively. Since targeted management actions are lacking in many localities integration of the strategies along with ongoing monitoring stakeholder engagement and dissemination for native species provides the most viable means of addressing invasive Big-leaf Mahogany (S. macrophylla) behavior (Coracero, 2023).

Big-leaf Mahogany (S. macrophylla) shoot borers have been widely studied, but no single control method is consistently effective, making integrated pest management essential (Newton et al., 1993). Mechanical and cultural practices such as mixed-species planting, wider spacing, enrichment planting, and providing lateral shade can reduce attacks by limiting host detection and improving tree vigor, though results vary by site conditions. Biological controls, including more than 50 recorded parasitoids, predators, and pathogens, have shown limited field success because larvae remain protected inside shoots. Chemical controls are similarly constrained by heavy rainfall, environmental concerns, and low efficacy, with both foliar and systemic insecticides proving unreliable across locations. Given these limitations, recommended management emphasizes combining mechanical, biological, and chemical methods, alongside breeding for resistant stock and designing plantation layouts that naturally.

5. Original Field Photos

Fig. 3 Big-leaf Mahogany (*S. macrophylla*) trees planted as living fences

Fig. 4 Big-leaf Mahogany (S. *macrophylla*) crown with immature fruit

Fig. 5 Leaf morphology of Big-leaf Mahogany (*S. macrophylla*). The leaf composition, phyllotaxy (top left), terminal bud (top right), adaxial leaf (bottom left), and abaxial leaf (bottom right)

Fig. 6 Big-leaf Mahogany (*S. macrophylla*) sapling

Fig. 7 Big-leaf Mahogany (*S. macrophylla*) trunk

Invasive Species Concept Map

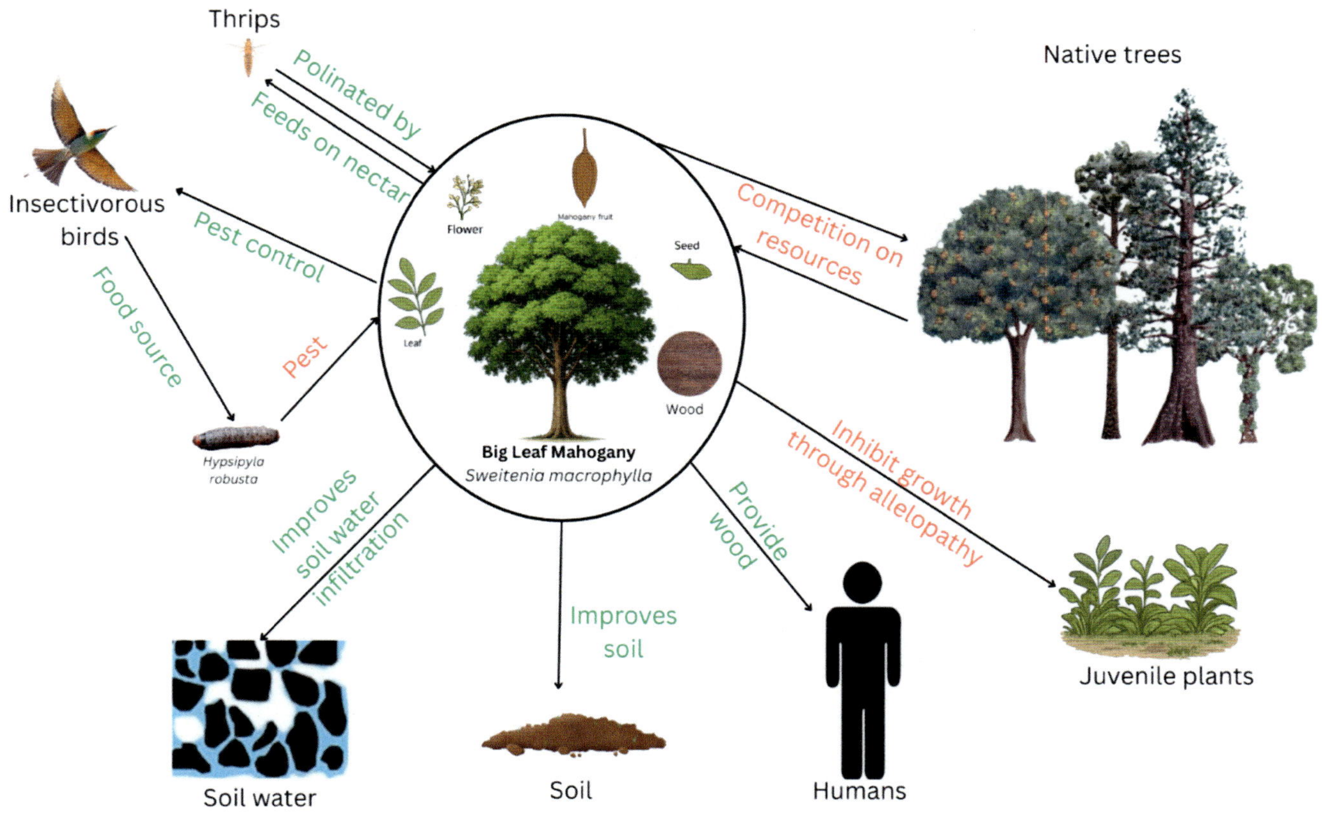

Big-leaf Mahogany (*S. macrophylla*), commonly referred to as bigleaf mahogany or Honduras mahogany, is a prominent tropical hardwood species native to Central and South America. It is known for its considerable height, often reaching up to 30 meters, and trunk diameters of approximately 1.5 meters. The species was introduced to the Philippines in 1907, initially cultivated as ornamental park trees in Manila. Due to its ecological adaptability and commercial value, S. macrophylla has become a subject of interest in forest ecology and management.

Big-leaf Mahogany (*S. macrophylla*) trees contribute significantly to soil enrichment through the decomposition of leaf litter, which enhances nutrient cycling and increases organic matter content. Their extensive root systems allow for efficient utilization of soil water. However, the large biomass of Big-leaf Mahogany (*S. macrophylla*) can lead to competitive interactions with surrounding vegetation, potentially reducing water availability for neighboring plant species (Grogan et al., 2014).

Big-leaf Mahogany (*S. macrophylla*) exhibits allelopathic properties, releasing biochemical compounds that inhibit the growth of juvenile plants in its vicinity. This phenomenon provides a competitive advantage but may suppress understory biodiversity. In mixed-species plantations or disturbed forest ecosystems, mahogany often outcompetes native tree species for essential resources such as light, nutrients, and water (CABI, 2023).

One of the most significant ecological challenges facing Big-leaf Mahogany (*S. macrophylla*) is infestation by the Hypsipyla shootborer (*Hypsipyla robusta*). This insect feeds on the shoots and fruits of mahogany, disrupting reproductive processes and stunting growth. The pest-host relationship is a critical factor in mahogany population dynamics and has implications for sustainable forest management (USDA Forest Service, 2015). This insect can be eaten by insectivorous animals, such as birds, lizards, and other arthropods.

The reproductive ecology of Big-leaf Mahogany (*S. macrophylla*) is characterized by its self-compatible flowers, which are primarily pollinated and may also be inhabited by thrips. The wingless immature stages of thrips dominate the population on mahogany trees and are most effective in facilitating within-tree pollination. Cross-pollination is less frequent due to the scarcity of pistillate flowers and winged adult thrips (Grogan & Loveless, 2013). Mahogany reproduces via seeds enclosed in large woody fruits, with successful dispersal and germination influenced by environmental conditions and pest pressure.

Big-leaf Mahogany (*S. macrophylla*) has several roles in the environment which is more efficient in the ecosystem dynamics of its native habitat. It can enhance soil enhancement, regulate water and nutrient cycling, provide wood to humans, and provide habitats and food sources to other species. In some environments Mahogany may do more harm than good as it is very good in competing for resources with other vegetation, as well as suppressing other juvenile plants through allelopathy and allowing only its regenerants to grow effectively.

In conclusion, Big-leaf Mahogany (*S. macrophylla*) plays a multifaceted role in forest ecosystems, functioning both as a valuable ecological contributor and a potentially dominant competitor. Therefore, while Big-leaf Mahogany (*S. macrophylla*) offers substantial benefits, its cultivation and integration into forest systems require careful ecological assessment and sustainable management strategies to balance productivity with biodiversity conservation.

References

Abarquez, A., Bush, D., Ata, J., Tolentino, E. Jr., & Gilbero, D. (2015). Early growth and genetic variation of mahogany (*Swietenia macrophylla*) in progeny tests planted in northern Mindanao, Philippines. Journal of Tropical Forest Science, 27(3), 314–324.

Baguinon, N. T., Quimado, M. O., & Francisco, G. J. (2005). Country report on forest invasive species in the Philippines. In Proceedings of the Asia-Pacific Forest Invasive Conference (17–23 August 2003). Kunming, China.

Bakewell-Stone, P. (2022). *Swietenia macrophylla* (big-leaved mahogany). CABI Compendium. https://doi.org/10.1079/cabicompendium.52155

Barstow, M. & Negrão, R. 2023. *Swietenia macrophylla*. The IUCN Red List of Threatened Species 2023: e.T32293A68104718. Retrieved from: https://dx.doi.org/10.2305/IUCN.UK.2023-1.RLTS.T32293A68104718.en. Accessed on 28 November 2025.

Basu, A., Ghosh, A., Pal, S., Basu, M., Karmakar, P., Mitra, B., & Pal, P. (2013). Pollination of the mahogany trees (*Swietenia mahagoni*) by an ovipositing thysanopteran insect (Priesneriola). [Unpublished manuscript].

Chudnoff, M. (1984). *Swietenia macrophylla*. In Tropical timbers of the world (USDA Forest Service, Agricultural Handbook No. 607). U.S. Department of Agriculture, Forest Products Laboratory. Retrieved from https://www.fpl.fs.usda.gov/documnts/TechSheets/Chudnoff/TropAmerican/pdf_files/swiete1new.pdf

Coracero, E. E. (2023). Distribution and Management of the Invasive *Swietenia macrophylla* King (Meliaceae) at the Foot of a Protected Area in Luzon Island, Philippines. Journal of Zoological and Botanical Gardens, 4(3), 637-647. https://doi.org/10.3390/jzbg4030045

Cossid, R. N., Torralba, J. M. A., Clemente, J. H. M., Villaflor, C. J. G., Jandug, C. M. B., & Casilac Jr, C. S. (2025). Bending Strengths of Large-leaf Mahogany (*Swietenia macrophylla* King) and Mangium (Acacia mangium Willd) Commercial Lumbers in Northeastern Mindanao, Philippines. Jurnal Sylva Lestari, 13(1), 317–331. https://doi.org/10.23960/jsl.v13i1.1072

Creencia, Glenn Bryan & Sarmiento, Daniel & Jr, Jose. (2021). Storage Potential of Mahogany (*Swietenia* spp.) Forest in Cavite State University, Philippines: A Basis for Carbon Mitigation Plan. 1-12.

Easyscape. (n.d.). Big Leaf Mahogany (*Swietenia macrophylla*). Retrieved [date you accessed it], from https://easyscape.com/species/Swietenia macrophylla%28Big Leaf Mahogany%29

FAO. (n.d.). *Swietenia macrophylla*. In Trees commonly cultivated in Southeast Asia (Section: Meliaceae). Retrieved [date you accessed], from https://www.fao.org/4/AC775E/AC775E04.htm

Food and Agriculture Organization of the United Nations. (n.d.). [Title of the specific FAO document / chapter on Swietenia macrophylla]. In FAO Forestry / FAO publication series. Retrieved [access date], from https://www.fao.org/4/ae944e/ae944e09.htm

Genovesi, P., & Monaco, A. (2013). Guidelines for addressing invasive species in protected areas. In L. C. Foxcroft, P. Pyšek, D. M. Richardson, & P. Genovesi (Eds.), Plant invasions in protected areas: Patterns, problems and challenges (pp. 487–506). Springer.

Grogan, J., Landis, R. M., Free, C. M., Schulze, M. D., Lentini, M., & Ashton, M. S. (2014). Big-leaf mahogany (*Swietenia macrophylla*) population dynamics and implications for sustainable management. Journal of Applied Ecology, 51(3), 664–674. https://doi.org/10.1111/1365-2664.12210

Grogan, James & Barreto, Paulo & Veríssimo, Adalberto. (2002). Mahogany in the Brazilian Amazon: Ecology and Perspectives on Management. 10.13140/RG.2.1.5054.6089.

Grogan, J. (2008). Mahogany (Swietenia macrophylla King). In Big-leaf Mahogany in the Brazilian Amazon: Ecology and Management (book chapter). CITES. Retrieved from CITES website

Howard, F. W., Nakahara, S., & Williams, D. S. (1995). Thysanoptera as apparent pollinators of West Indies mahogany, *Swietenia mahagoni* (Meliaceae). Annales des Sciences Forestières, 52(3), 283–286. https://hal.archives-ouvertes.fr/hal-00882998

Joshi, R. C. (2006). Invasive alien species (IAS): Concerns and status in the Philippines. Philippine Rice Research Institute.

Newton, A. C., Baker, P., Ramnarine, S., Mesén, J. F., & Leakey, R. R. B. (1993). The mahogany shoot borer: Prospects for control. Forest Ecology and Management, 57(1–4), 301–328. https://doi.org/10.1016/0378-1127(93)90179-Q

Norghauer, J. M., Malcolm, J. R., & Zimmerman, B. L. (2006). Juvenile mortality and attacks by a specialist herbivore increase with conspecific adult basal area of Amazonian *Swietenia macrophylla* (Meliaceae). Journal of Tropical Ecology, 22(4), 451–460. https://doi.org/10.1017/S0266467406003303

Orwa, C., Mutua, A., Kindt, R., Jamnadass, R., & Simons, A. (2009). *Swietenia macrophylla* King (Big-leaf mahogany) [Agroforestry Database 4.0]. World Agroforestry Centre. https://apps.worldagroforestry.org/treedb2/AFTPDFS/Swietenia_macrophylla.PDF

Patiño Valera, F., Centeno Erguera, R., & Marín Chávez, J. (2003). Conservation and use of mahogany in forest ecosystems in Mexico (Case Study No. 6). In Biodiversity and the ecosystem approach in agriculture, forestry and fisheries. Food and Agriculture Organization of the United Nations. https://www.fao.org/4/Y4586E/y4586e07.htm

Tulod, A. M., Casas, J. V., Marin, R. A., & Ejoc, J. A. B. (2017). Diversity of native woody regeneration in exotic tree plantations and natural forest in Southern Philippines. Forest Science and Technology, 13(1), 31–40. https://doi.org/10.1080/21580103.2017.1292958

Ipil-Ipil (*Leucaena leucocephala*): Impacts and Control

Faith Ezaira T. Ponciano & Hazel Grace B. Yacon

I. Species History, Information, and Characteristics

Common Name: Ipil-ipil

Scientific Name: *Leucaena leucocephala* (Lam.) De Wit

Family Name: Fabaceae

A. Native Range and Method of Introduction

Leucaena leucocephala (Lam.) De Wit is an indigenous tree of the Yucatan Peninsula in southern Mexico. This tree generally grows in tropical and subtropical eco-climatic zones far from marginal ice zones. For optimal growth of the tree, warm temperatures of about 25–30°C are required. The *L. leucocephala* tree can be seen on the edges of roads, forests, and in suburban areas where it has been planted. However, at present, it has spread to different parts of the world, which are usually found in Central America, Asia, Africa, and Australia, with small distributions in Europe, and it is not found in Atlantic countries. (Ilham et al., 2015)

Among 700 international collections of this self-fertile species grown in Hawaii, most of them were identical seedy shrubs, and in the warm tropics, the 'common type' was widely used for fodder and fuelwood. In contrast, seed collections in their native Mexico and Central America range widely in ideotype, which includes arboreal types from which 'giant' cultivars have been bred and are now internationally in use. (Brewbaker, 2004)

Lower rate of spread under grazing, as cattle relish young seedlings. However, if it is left unmanaged during its growing season, stand density can increase greatly due to seed drop or by wind dispersal. Another seed can be further disseminated by passage through the grazing animal or by water movement. Thickening up of hedgerows may occur where delayed grazing allows seed set and seedling recruitment.

Ipil-ipil (*L. leucocephala*) is native to Central America and Mexico, where it naturally occurs in dry tropical forests and grasslands. Due to its favorable traits, such as rapid growth, high nitrogen-fixing capacity, and tolerance to poor soils, it was intentionally introduced across Asia, including the Philippines. The species was historically brought into the country primarily for reforestation, erosion control, fuelwood production, and livestock fodder, which was under the early 20th-century forestry and agricultural programs. Ipil-ipil (*L. leucocephala*) was promoted by the government agencies and development projects because of its ability to rehabilitate the degraded lands and provide a readily available source of biomass and forage. Over time, however, it became naturalized and spread beyond its cultivation, resulting in both ecological benefits and concerns, such as Ipil-ipil's competitiveness with the native species. (Brewbaker, 1987).

B. Ecological and/or Economic Impact

Ecological

Ipil-ipil (*L. leucocephala*) has significant ecological impacts in invaded habitats, largely due to its rapid growth and ability to outcompete the native species in the area. It forms dense monospecific stands that suppress neighboring woody and herbaceous vegetation through both competition for light and allelopathic effects; its leaf litter has been shown to inhibit the germination and growth of a variety of species (Sharma et al., 2022).

Allelopathy

Allelopathy is a chemical interaction among plants caused by the allelochemicals (Rice, 1984), which are produced in plants and released into the vicinity of the plants, including rhizosphere soil, either by root exudation, decomposition of plant litter and residues, rainfall leachates, or volatilization from the plant parts. It is the substances that inhibit the germination, the growth, and the survival of the nearby plants, like weeds, woody species, and native trees, that greatly affect them (Belz, 2007).

Plant Litter and Residue

Leaf litter of *L. leucocephala* was mixed with the soil and the seeds of a woody plant, *Albizia procera* (Roxb.) Benth., and crop plants, *Vigna unguiculata* (L.) Walp., and *Cicer arietinum* L., and *Cajanus cajan* (L.) Millsp. were sown into the mixture. The treatments resulted in the suppression of the germination and growth of these test plant species (Ahmed et al., 2008). Soil mixture with decomposing leaves of *L. Leucocephala* increased the mortality of five tree species: *Alnus formosana* (Burkill) Makino, *Acacia confusa* Marr., *Liquidambar formosana* Hance, *Casuarina glauca* Sieber, and *Mimosa pudica* L. (Chou & Kuo, 1986).

Aqueous extracts of *L. Leucocephala* litter, which accumulated on the forest floors, showed the suppression of the germination and radicle growth of Lolium multiflorum Lam. (Chou & Kuo, 1986), and *Ageratum conyzoides* L., *Tridax procumbens* L., and *Emilia sonchifolia* (L.) DC. ex Wight (Ishak et al., 2026). Leaf mulch of *L. Leucocephala* covered on the soil surface or mixed with soil inhibited the germination and growth of *Vigna unguiculata* (L.) Walp. and the root nodulation of *V. unguiculata* (Kamara et al., 1999). Those observations indicate that the leaf litter and residues of *L. Leucocephala* may contain certain allelochemicals, and some of them may be liberated into the soil during their decomposition processes (Kato-Noguchi & Kurniadie, 2022).

Its high capacity for nitrogen fixation elevates its soil nitrogen levels, which can alter the nutrient cycling and favor the other invasive species, thereby disrupting community structure (Sun et al., 2025).

The reproductive and life-history traits, including abundant seed production, resprouting ability, and release of defensive compounds such as mimosine, further promote its invasiveness. Moreover, the invasion of *L. Leucocephala* can alter forest succession. In regenerating forest fragments, its dominance reduces species richness and functional diversity, especially among the animal-dispersed species, potentially compromising long-term ecosystem recovery (Santos et al., 2023).

Finally, its ecological impacts extend to landscape-level processes because it tolerates the disturbance and resprouts after a fire or cutting. *L. Leucocephala* can maintain its dominance, reshaping the vegetation patterns, succession trajectories, and the overall assembly of plant communities. (Sharma et al., 2022).

Chemical composition and nutritive value:

Leucaena leucocephala leaves and seeds contain lipids, crude protein, and carbohydrates. The seeds contain tannin and oxalic acid (Gutteridge et al., 1998). The kernel contains an oil content of about 17-20% (Hossain et al., 1998). The leaves and seeds also contain a toxic and non-protein substance known as mimosine.

Economic

Agricultural Purpose

The interesting part and contribution of *L. Leucocephala* is linked to its versatility, which makes it a multipurpose plant that can provide several usage options. It is a soil-improving plant species. (Halliday et al., 2014). It can be used as a living fence around the garden for the beautification of the landscape, a fire break, and a windbreak in the forest area (Meena Devi et al., 2013). Further, it can be a source of pharmacological compounds for veterinary applications, and a forage plant for honeybees and other pollinators (Damascena et al., 2017).

Biofuel

L. leucocephala is an excellent firewood species with a specific gravity and high calorific value (Whitesell, 1974). Wood burns steadily with little smoke and few sparks and produces less than 1% ash. The tree makes excellent charcoal. *L. Leucocephala* seed oil can be used as an oil extract for biofuel production (Yusuff et al., 2019) and is used as a biofuel in diesel engines.

Dyes

Red, brown, and black dyes are extracted from the pods, leaves, and bark of *L. Leucocephala*. It contains tannin and is used as a natural dye in the leather and cotton industries (Shrivastava, 2012).

Gums

Gum extracted from *L. Leucocephala* stems is used as a gum and has potential commercial value. Seeds also yield about 25% gum and form highly viscous solutions at low solute concentrations (Raval et al., 1988).

Human food

L. Leucocephala is consumed by humans in Central America, Indonesia, and Thailand. It can be eaten in processed and unprocessed forms. In Java, the seeds are fermented and are eaten as sprouts or bean cake. In the Philippines, the young pods are cooked as a vegetable, and roasted seeds are used as a substitute for coffee. The young dry seeds are popped like popcorn (Hegarty et al., 1964). In Indonesia, Thailand, Mexico, and Central America, people also eat the young leaves, flowers, and young pods made into soups and salads. Seeds are being considered as a non-conventional source of protein, together with other leguminous seeds (Azeemoddin et al., 1988).

Medicinal uses

The seeds of *L. Leucocephala* have great medicinal properties and are used to control stomach aches and as contraception and abortifacients. The seed gum can be used as a binder in tablet formulation (Verma and Balkishen, 2007). Sulfated glycosylated forms of polysaccharides from the seeds were reported to possess significant cancer chemopreventive and anti-proliferative activities (Gamal-Eldeen, 2007). The extracts of the seeds have been reported as anthelmintic and antidiabetic and have a broad spectrum of antibacterial activity (Ademola, 2005). Recently, the seed oil was used in engineering as a novel bio-device useful in biomembrane modeling in the lipophilicity determination of drugs and xenobiotics (Idowu, 2009). The plant is reported to be a worm repellent.

Timber

L. Leucocephala wood is pale yellow sapwood and light reddish-brown heartwood. The wood has a medium density, a medium texture, and a close grain and is easily workable for a wide variety of carpentry purposes like sawn timber, mine props, furniture, and parquet flooring (Alabi and Alausa, 2006). *L. leucocephala* pulp is used in paper and rayon industries (Dutt et al., 2007).

C. Control and Eradication Measures

a. Mechanical Control

Mowing and hand weeding have shown a higher targeted effectiveness in field invasion control, with limitations such as high labor costs and frequently required repeated treatments. Cutting trees close to the ground, uprooting small plants, hand-pulling seedlings, and repeated slashing or mowing for regrowth suppression. Girdling, or trunk-ringing, is used to starve large stems. For multi-stem or large trees, mechanical removal is usually followed immediately by the stump treatment (chemical) to prevent its resprouting. In addition, it can also be used as soil covering or shading (plastic sheets) on the cut trunk to reduce sprouts in small sites (Flory and Lewis, 2009).

b. Biological Control

The invasive potential of *L. leucocephala* can be further controlled through a careful selection of more vigorous/less seed-producing genotypes. (Campbell et al., 2019). Biological control against *L. leucocephala* itself is limited—most classical biocontrol work has targeted the psyllid (Heteropsylla cubana), a sap-sucking insect that can cause severe defoliation of *L. leucocephala*, Bruchid, Acanthoscelides macrophthalmus, are also used in biocontrol of *L. leucocephala*. Soil insects such as earwigs, scarab beetles, termites, and cutworms can cause serious damage to emerging seedlings. A range of pathogenic fungi and insects occasionally attack *L. leucocephala*. Damping off diseases caused by the fungal species Pythium or Rhizoctonia commonly kill newly emerged nursery and field-grown seedlings.

c. Chemical Control

L. Leucocephala can be controlled by basal bark application of herbicides containing 120 g/L picloram and 240 g/L triclopyr mixed with diesel, aminopyralid + metsulfuron-methyl, clopyralid, glyphosate, imazapyr, metsulfuron-methyl, picloram, triclopyr + picloram, and triclopyr + picloram + aminopyralid. Application of glyphosate to regrowth after slashing will kill trees, but repeat applications may be necessary. The efficacy of herbicides applied using three drill-applied methods (stem implant, drill and fill, and basal bark) to control *L. leucocephala* with stems up to 46 cm in circumference. Capsules containing picloram, clopyralid, and metsulfuron + aminopyralid were as effective as the alternative reference treatments (Brazier et al., 2023).

II. Photos

(Pictures were originally captured at Brgy. Sampaloc II, Sariaya, Quezon, and Brgy. Talipan, Pagbilao, Quezon)

Fig. 1 Ipil-ipil (*L. Leucocephala*) branches showing the flower buds and flowers in blooms, and fruits (top) and close up view of matured fruits (bottom)

Fig. 2 Morphology of fully matured Ipil-ipil (*L. Leucocephala*) tree

Fig. 3 Close up view of trunk (top) and Ipil-ipil (*L. Leucocephala*) trees growing in the field (bottom)

III. Invasive Species Concept Map

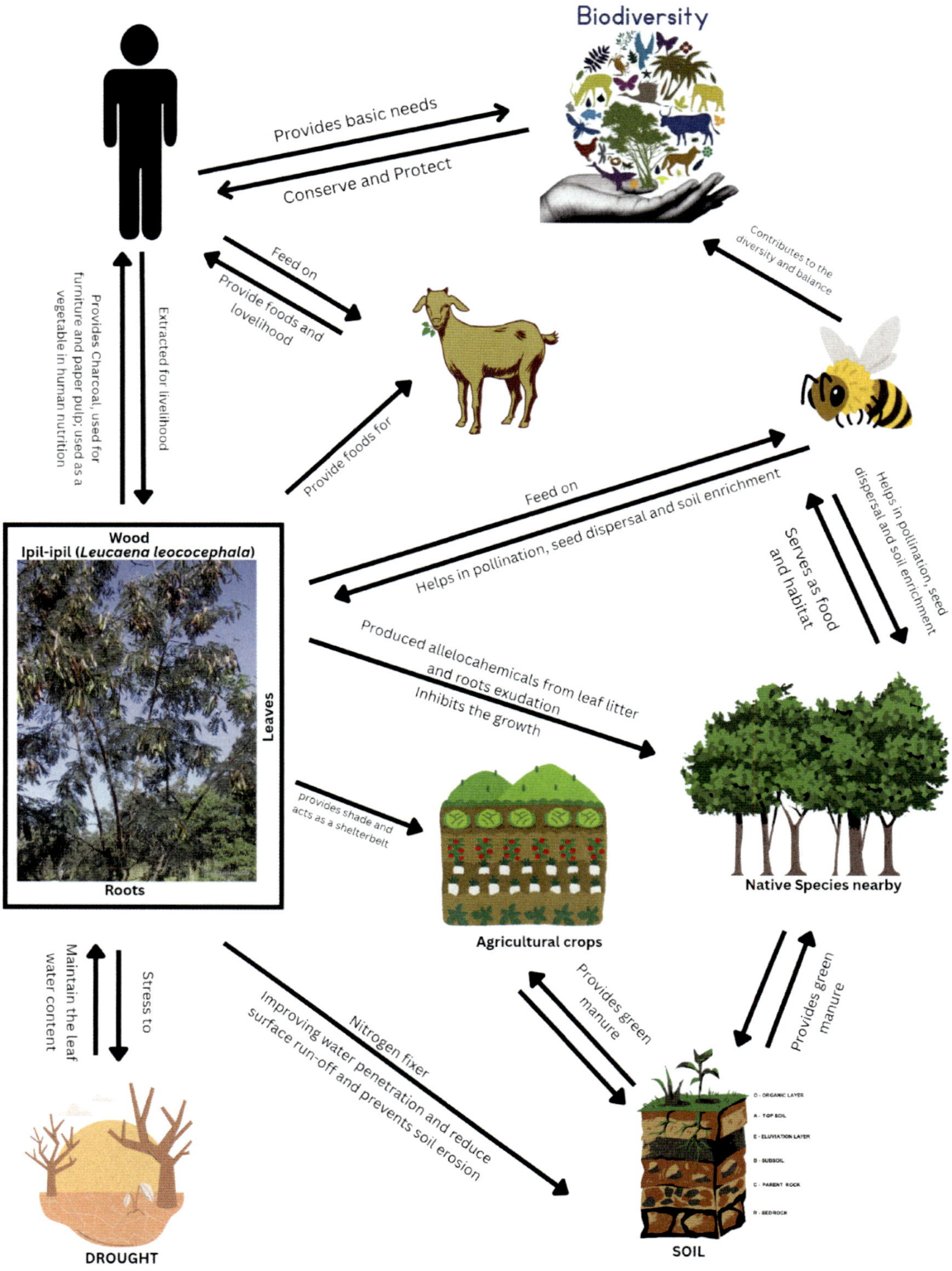

Human Interaction

Humans extract Ipil-Ipil (*L. leucocephala*) wood for charcoal, fuel, furniture, and other livelihood needs. This tree provides essential resources and supports income generation. At the same time, humans are also responsible for conserving and protecting biodiversity, ensuring that the use of Ipil-Ipil (*L. leucocephala*) remains sustainable. Its fast growth and multipurpose uses make it highly valuable, but uncontrolled spread can create ecological challenges.

Livestock (e.g., goat, cow, carabao)

Goats feed on Ipil-Ipil (*L. leucocephala*) leaves, gaining high-quality nutrition that supports livestock production. In return, livestock provide food and livelihood benefits to humans. The tree's nutrient-rich leaves make it a widely used fodder species, reinforcing its importance in small-scale farming systems.

Pollinators (e.g., bees and other insects)

Bees feed on the flowers of Ipil-Ipil (*Leucaena leucocephala*), helping with pollination. Fruits are dispersed by wind. Leaf litter decomposition contributes to soil enrichment. The activity supports ecosystem balance and biodiversity. Through attracting pollinators, Ipil-Ipil (*Leucaena leucocephala*) indirectly strengthens ecological linkages, though its spread must be monitored.

Ipil-ipil Tree (Wood, Leaves, Roots)

The diagram shows Ipil-Ipil (*L. leucocephala*) providing wood, fodder, shade, and environmental services such as soil improvement. However, the tree also releases allelochemicals from its leaves and roots, which can inhibit the growth of nearby crops and native plants. This dual role of being beneficial but also invasive captures Ipil-Ipil's (*L. leucocephala*) ecological complexity. It enriches soil but can suppress competing plants through allelopathy.

Agricultural Crops

Decomposed leaves and other plant parts of leguminous species are rich in nitrogen, making them valuable for improving soil fertility. Ipil-Ipil (*L. leucocephala*), in particular, provides multiple benefits to crops, including shade, wind protection, and enhanced soil quality—especially in degraded or exposed farmlands. However, the tree also produces allelochemicals that can inhibit crop germination and growth. This dual effect explains why Ipil-Ipil (*L. leucocephala*) is highly valued in agroforestry systems yet also considered a management challenge.

Native Plant Species

Native vegetation may receive indirect benefits from pollinators and improved soil conditions, but the allelopathic properties and fast growth of Ipil-Ipil (*L. leucocephala*) can inhibit or outcompete them. The concept map reflects that the species' invasive potential tends to dominate and reduce native plant diversity.

Soil

Ipil-Ipil (*L. leucocephala*) improves soil fertility through nitrogen fixation, green manure production, increased water infiltration, and reduced erosion. These functions make it highly useful for land restoration. These soil benefits explain why the species is widely planted, even though its competitive nature requires careful control.

Drought Conditions

Ipil-Ipil (*L. leucocephala*) during drought helps maintain soil moisture due to its deep roots. It can survive dry periods better than many species. Its drought tolerance contributes to its resilience and rapid spread, especially in degraded landscapes.

Biodiversity

Biodiversity benefits from pollination, soil enrichment, and ecological interactions. Humans play a role in maintaining this balance by utilizing Ipil-Ipil (*L. leucocephala*) sustainably. While this species supports certain ecosystem functions, its invasive tendencies can reduce overall biodiversity if unmanaged and uncontrolled.

References

Ademola, I. O., Akanbi, A. I., & Idowu, S. O. (2005). Anthelmintic activity of *Leucaena leucocephala* seed. Pharmaceutical Biology, 43(7), 599–604.

Ahmed, T., Hoque, A. T. M. R., & Hossain, M. K. (2008). Allelopathic effects of *Leucaena leucocephala* leaf litter on some forest and agricultural crops grown in nurseries. Journal of Forestry Research, 19(4), 298–302. https://doi.org/10.1007/s11676-008-0051-7

Alabi, D. A., & Alausa, A. A. (2006). Evaluation of the mineral and phytochemical contents of *Leucaena leucocephala*. World Journal of Agricultural Sciences, 2(1), 115–118.

Alemán-Ramirez, J. L., Okoye, P. U., Torres-Arellano, S., Mejía-Lopez, M., & Sebastian, P. J. (2022). A review on bioenergetic applications of *Leucaena leucocephala*. Industrial Crops and Products, 182, Article 114847. https://doi.org/10.1016/j.indcrop.2022.114847

Azeemoddin, G., Jagan Mohan Rao, S., & Thirumala Rao, S. D. (1988). Studies on the composition of *Leucaena leucocephala* oil. Journal of Food Science and Technology, 25, 158.

Brazier, D. A., Brooks, S. J., Warren, C., & Gough, K. L. (Year). CONTROLLING LEUCAENA LEUCOCEPHALA WITH ENCAPSULATED HERBICIDES. Tropical Weeds Research Centre, Department of Agriculture and Fisheries, Charters Towers, Queensland, Australia.

Belz, R. G. (2007). Allelopathy in crop/weed interactions—An update. Pest Management Science, 63(4), 308–326. https://doi.org/10.1002/ps.1347

Brewbaker, J. L. (1987). LEUCAENA: A MULTIPURPOSE TREE GENUS FOR TROPICAL AGROFORESTRY. In N. T. Vergara (Ed.), Agroforestry in the tropics (pp. 39–62). Southeast Asian Regional Center for Graduate Study and Research in Agriculture.

Brewbaker, J. L. (2004). Tree breeding practices: Nitrogen-fixing tree improvement and culture. In Encyclopedia of Forest Sciences. Retrieved from https://www.example.com/encyclopedia

Campbell, S., Vogler, W., Brazier, D., Vitelli, J., & Brooks, S. (2019). Weed Leucaena and its significance, implications, and control. Tropical Grasslands – Forrajes Tropicales, 7(3), 280–289. https://doi.org/10.17138/TGFT(7)280-289

Chen, Y., Chen, F., Liu, L., & Zhu, S. (2012). Physiological responses of *Leucaena leucocephala* seedlings to drought stress.

Chiou, C. R., Wang, H. H., Chen, Y. J., & Grant, W. E. (2013). Modeling potential range expansion of the invasive shrub *Leucaena leucocephala* in the Hengchun Peninsula, Taiwan. Invasive Plant Science and Management, 6(3), 492–501. https://doi.org/10.1614/IPSM-D-12-00057.1

Chou, C. H., & Kuo, Y. L. (1986). Allelopathic research of subtropical vegetation in Taiwan. Journal of Chemical Ecology, 12(6), 1431–1448. https://doi.org/10.1007/BF01012333

Damascena, J. G., Leite, G. L. D., Silva, F. W. S., Soares, M. A., Guanabens, R. E. M., Sampaio, R. A., & Zanuncio, J. C. (2017). Spatial distribution of phytophagous insects, natural enemies, and pollinators on the *Leucaena leucocephala* (Fabaceae) tree in the Cerrado. Florida Entomologist, 100(3), 571–577. https://doi.org/10.1653/024.100.0303

De Angelis, A., Gasco, L., Parisi, G., & Danieli, P. P. (2021). A multipurpose leguminous plant for the Mediterranean countries: *Leucaena leucocephala* as an alternative protein source: A review. Animals, 11(8), 2230. https://doi.org/10.3390/ani11082230

Dutt, D., Tyagi, C. H., & Malik, R. S. (2007). Chemical technology evaluation of *Leucaena leucocephala*. Indian Journal of Chemical Technology, 14, 626–634.

(FAO, 1983) FAO, 1983. Handbook on Taxonomy of Prosopis in Mexico, Peru, and Chile. Available. in: http://www.fao.org/documents/endetail/144974

Flory, SL, Lewis, J. 2009. Nonchemical methods for managing Japanese stiltgrass (MICROSTEGIUM VIMINEUM). INVASIVE PLANT SCIENCE AND MANAGEMENT 2(4): 301–308.

Gamal-Eldeen, A. M., Amer, H., Helmy, W. A., Ragab, H. M., & Talaat, R. M. (2007). Biological properties of *Leucaena leucocephala* constituents. Indian Journal of Pharmaceutical Sciences, 69, 805–811.

Gutteridge, R. C., & Shelton, H. M. (1998). *Leucaena* as forage: A review. Tropical Grassland Society of Australia, 2, 1.

Halliday, M. J., Panjaitan, T., Nulik, J., Padmanabha, J., McSweeney, C. S., Depamede, S., KanaHau, D. K., Fauzan, M., Yuliana, B. T., Pakereng, C. (2014). Prevalence of DHP toxicity and detection of Synergistes jonesii in ruminants consuming Leucaena leucocephala in eastern Indonesia. Tropical Grasslands – Forrajes Tropicales, 2(1), 71–73. https://doi.org/10.17138/TGFT(2)71-73

Hegarty, M. P., Schinckel, P. G., & Court, R. D. (1964). Toxic factors in *Leucaena leucocephala*. Australian Journal of Agricultural Research, 15, 153–167.

Hossain, M. A., Alam, M., & Huq, M. S. (1998). Studies on the composition of Ipil-ipil (*Leucaena leucocephala*) seed oil. Dhaka University Studies, Part B, 36, 163–169.

Idowu, S. O., Adeyemo, M. A., & Ogbonna, U. I. (2009). Biological evaluation of *Leucaena* extracts. Journal of Biological Engineering, 3, 14.

Ilham, Z. (2015). Extraction and quantification of the toxic compound mimosine from *Leucaena leucocephala* leaves. Procedia Chemistry, 16, 164-170. https://doi.org/10.1016/j.proche.2015.12.029

Ishak, M. S., Ismail, B. S., & Yusoff, N. (2016). Allelopathic potential of *Leucaena leucocephala* (Lam.) de Wit on the germination and seedling growth of Ageratum conyzoides L., Tridax procumbens L., and Emilia sonchifolia (L.) DC. Allelopathy Journal, 37(1), 109–122.

Kamara, A. Y., Akobundu, I. O., Sanginga, N., & Jutzi, S. C. (1999). Effects of mulch from 14 multipurpose tree species (MPTs) on early growth and nodulation of cowpea (Vigna unguiculata L.). Journal of Agronomy and Crop Science, 182(2), 127–133. https://doi.org/10.1046/j.1439-037x.1999.00278.x

Kato-Noguchi, H., & Kurniadie, D. (2022). Allelopathy and allelochemicals of *Leucaena leucocephala* as an invasive plant species. Plants, 11(13), 1672. https://doi.org/10.3390/plants11131672

Meena Devi, V. N., Ariharan, V. N., & Nagendra Prasad, P. (2013). Nutritive value and potential uses of *Leucaena leucocephala* as biofuel—A mini review. Research Journal of Pharmaceutical, Biological and Chemical Sciences, 4(1), 5515–5521.

Njunge, J. T., Kaholongo, I. K., & Amutenya, M. (2017). Invasiveness and biomass production of *Leucaena leucocephala* under harsh ecological conditions of north-central Namibia. Journal of Tropical Forest Science, 29(3), 297–304.

Ojo, O. A., & Fagade, O. E. (2002). Nutritional potentials of *Leucaena leucocephala*. African Journal of Biotechnology, 1(1), 23–27.

Raval, D. K., Patel, R. G., & Patel, V. S. (1988). Starch properties of *Leucaena leucocephala*. Starch, 40, 214–218.

Rice, E. L. (1984). Allelopathy (2nd ed.). Academic Press.

Santos, B. A., Peres, C. A., & Benchimol, M. (2023). FOREST REGENERATION UNDER INVASION: IMPACTS OF *Leucaena leucocephala* ON SPECIES RICHNESS AND FUNCTIONAL DIVERSITY. EcoEvoRxiv. https://doi.org/10.32942/osf.io/xxxxx

SHARMA, P., KAUR, A., BATISH, D. R., KAUR, S., & CHAUHAN, B. S. (2022). CRITICAL INSIGHTS INTO THE ECOLOGICAL AND INVASIVE ATTRIBUTES OF LEUCAENA LEUCOCEPHALA, A TROPICAL AGROFORESTRY SPECIES. Frontiers in Agronomy, 4, 890992. https://doi.org/10.3389/fagro.2022.890992

Shrivastava, V. S. (2012). Chemical studies on *Leucaena leucocephala*. International Journal of ChemTech Research, 4(3), 1038–1043.

Sun, Y., Li, Z., Ma, J., Qiao, L., Liu, Y., & Li, S. (2025). PHYTOCHEMICAL ANALYSIS AND ALLELOPATHIC EFFECTS OF *Leucaena leucocephala* LEAF EXTRACTS. Molecules, 30(11), 2453. https://doi.org/10.3390/molecules30112453

Verma, P. R. P., & Balkishen, R. (2007). Physicochemical studies on *Leucaena leucocephala*. Journal of Scientific and Industrial Research, 66, 550–557.

Whitesell, C. D. (1974). *Leucaena leucocephala*. In C. S. Schopmeyer (Ed.), Seeds of woody plants in the United States (USDA Forest Service Handbook No. 450). Washington, DC.

Witt, A. B. R. (2010). Biofuels and invasive species from an African perspective—A review. GCB Bioenergy, 2(6), 321–329. https://doi.org/10.1111/j.1757-1707.2010.01061.x

Yusuff, A. S., Lala, M. A., Popoola, L. T., & Adesina, O. A. (2019). Optimization of oil extraction from *Leucaena leucocephala* seed as an alternative low-grade feedstock for biodiesel production. SN Applied Sciences

C

University of the Philippines Los Baños College of Forestry and Natural Resources

Managing Water Hyacinth (*Pontederia crassipes*): Problems and Solutions

Elisha Ruth E. Sarita & Norgee B. Parreño

Species History, Information, and Characteristics

This section highlights one of the most productive plants on Earth, capable of yielding over 200 tons of dry matter per hectare annually under normal conditions. Commonly known as water hyacinth, *Pontederia crassipes* Mart. is classified as an invasive species. It was previously known as *Eichhornia crassipes* but was reclassified in 2018 as *Pontederia crassipes*.

Native Range and Introduction to the Philippines

Water hyacinth is a perennial, free-floating, and flowering aquatic plant that originated from the Amazon Basin, South America (Hill and Coetcze, 2008). It thrives in various freshwater ecosystems, including rivers, lakes, canals, and ponds (feedipedia.org).

Because of its striking lavender blooms, *Pontederia crassipes* was introduced in the Philippines as an ornamental for garden ponds (Francisco, 2020). This species has now infested major fresh water bodies such as the Pasig River, Laguna de Bay and Rio Grande de Mindanao and has been considered as part of the top 10 worst weeds in the world (Basilio & Buot, 2023) and an EPPO A2 Pest with strong efforts on its eradication and quarantine to prevent further spread to neighbouring countries (Eppo.int., 2019).

Ecological and Economic Impacts

Water hyacinth (*Pontederia crassipes*) is a prolific invasive species capable of reproducing both asexually and sexually. Through sexual reproduction, a single flower can produce up to 300 seeds, which remain viable for up to 20 years (Barrett, 1980). A single plant is also capable of producing more than 3,000 daughter plants within two months. This extraordinary reproductive capacity enables the species to proliferate rapidly, forming dense mats that drastically reduce dissolved oxygen levels and thereby threaten aquatic organisms. In addition, these mats obstruct waterways, impeding boating, sailing, and fishing activities (Francisco, 2020), effectively disrupting the local economy. Francisco (2020) further emphasized that unchecked proliferation of water hyacinth can lead to ecological devastation, as the mats not only deplete oxygen but also block sunlight, displacing or destroying native aquatic vegetation.

Despite its destructive potential, water hyacinth has demonstrated utility in environmental remediation. Its extensive root system can absorb and sequester minerals, heavy metals, and even radioactive elements from contaminated waters (Hasan et al., 2009a). Studies have shown its effectiveness in removing cadmium and zinc from moderately polluted water bodies (Lu XiaoMei et al., 2004).

Economically, water hyacinth has been integrated into innovative farming systems. For example, fish–pig–water hyacinth farming models have been developed to enhance global animal production. In such systems, water hyacinths cultivated in fishponds exhibit higher nutritive value and can be used as feed for fish and pigs, while animal manure fertilizes the ponds and supports hyacinth growth (Yang Huazhu et al., 2001). Beyond farming, water hyacinth has been utilized in textile and paper production, as camouflage for fish traps (Pieterse, 1997), and as biomass for biogas generation due to its rapid fermentation. It also serves as a water-clearing agent, a substrate for mushroom cultivation, and an ornamental plant (Pieterse, 1997). Owing to its high crude protein content, it has been employed as feed for ruminants, pigs, geese, ducks, and fish (Wimalarathne & Perera, 2019). More recently, it has been harnessed for bioenergy, briquettes, fertilizer, and even arts and crafts, and has also been introduced into educational curricula from elementary to high school (Harun et al., 2021).

Nevertheless, the economic drawbacks of water hyacinth are substantial. Its proliferation disrupts water-based activities such as fishing, irrigation, transportation, and hydroelectric operations, while also degrading water quality (Ecoport, 2011; Practical Action, 2006; Hasan et al., 2009b). Although conversion into compost and mulch can enhance agricultural yields and reduce feed costs (Jha et al., 2024), the plant's invasive nature continues to impose severe challenges. In the Philippines, its pervasive presence has obstructed waterways and hindered the fishing industry. During Typhoon Frank in 2008, Cotabato City reported that water hyacinth exacerbated flooding, affecting more than 5,000 households in nearby municipalities (jacintoandlirio.com). Furthermore, its clogging of irrigation channels and interference with power generation highlight its capacity to inflict significant economic losses.

Control and Eradication Measures

There have been multiple attempts to control the proliferation of water hyacinths; to some extent, eradication has been considered. Note that eradication is often difficult to achieve, and utilization has become an alternative pathway. *P. crassipes* has been used as animal feed, biofuel, fibre, textile, fertilizer and as phytoremediation for water pollution. Further research has been conducted to study the plant's composition and its medicinal value (Basilio & Buot, 2023).

Mechanical / Physical controls

Water hyacinth (*P. crassipes*) is recognized as one of the most notorious invasive species, having spread to more than 65 countries across Asia. Mechanical approaches are commonly employed for their management, as these methods do not degrade the environment or pollute freshwater ecosystems. While manual removal by hand may appear practical, the extensive root system and substantial biomass of water hyacinth often overwhelm available manpower. Moreover, manual extraction alone is insufficient to outpace the plant's rapid growth rate.

In cases of extensive infestation, the use of heavy equipment such as excavators, tractors, and boats becomes indispensable, emphasizing the importance of mechanized strategies in effective large-scale management. For instance, in 2012, the Department of Science and Technology's Metals Industry Research and Development Center (DOST-MIRDC) in the Philippines developed the Water Hyacinth Harvester, a machine capable of collecting up to 25 kilograms of water hyacinth per load.

China, a country heavily affected by water hyacinth, has adopted an integrated management strategy that combines multiple practices to both control and utilize the weed. Physical harvesting methods include dredging, mechanical harvesting, and the installation of floating booms and barriers. Manual collection is also employed for small-scale infestations, typically involving labor-intensive hand removal or the use of tools such as rakes and nets. For large-scale infestations, mechanical harvesting is preferred, utilizing specialized equipment mounted on boats with rotating blades or conveyor systems to cut and remove plants efficiently. This approach has proven more effective than manual methods (Jha et al., 2024).

In addition to mechanical strategies, arthropods have been introduced as biological regulators of water hyacinth populations. Chemical control methods, including the application of glyphosate, diquat, amitrole, and 2,4-D amine, are also used to suppress infestations. While manual extraction can be more effective than chemical or biological methods in small areas, it is impractical for large infestations due to the high labor demands and associated risks, particularly in regions where aquatic fauna such as snakes and crocodiles pose hazards to workers (Gupta & Yadav, 2020).

Biological controls

Day, et al (2023) reported for the first time the successful establishment of weevils released as biological control agents against water hyacinth (*P. crassipes*) in the Philippines. The two species, *Neochetina bruchi* and *N. eichhorniae*, were originally introduced in 1992 to suppress P. crassipes populations, but their establishment in the country had not previously been confirmed (Julien, 2001; Winston et al., 2014). The impact of these weevils on *P. crassipes* varies across regions, ranging from slight to substantial reductions, with more effective control achieved when both species are present. For example, in Papua New Guinea, biological control programs have demonstrated particularly successful outcomes at certain sites (Julien & Orapa, 2000; Day & Bule, 2016).

Historical evidence also supports their efficacy. A report from Sudan (1979) documented a 22.5% reduction in plant growth when progeny of *N. bruchi* and *N. eichhorniae* were cultured together, alongside a 136.6% increase in population control. Comparative trials further revealed that *N. bruchi* was more efficient *than N. eichhorniae* in suppressing hyacinth growth when stocked separately with adults and larvae (Bashir, 1984).

Beyond these two species, other insects have been identified as biological control agents of water hyacinth. The weevil *Sameodes albiguttalis* functions as a stem borer, while the semi-aquatic grasshopper *Cornops aquaticum* has been employed in South Africa. The latter not only feeds on water hyacinth but also facilitates secondary pathogenic infestations, as documented by Amedegnato (2008).

Chemical controls

The application of chemical herbicides is considered effective for managing extensive infestations of aquatic weeds; however, their use poses risks of pollution within freshwater ecosystems. Commonly employed herbicides include Bispyribac, Diquat, Glyphosate, Imazamox, Imazapyr, Penoxsulam, Triclopyr, 2,4-D, and Florpyrauxifen-benzyl, which have demonstrated notable efficacy in reducing populations of water hyacinth. Among these, Bispyribac, Glyphosate, Imazamox, Imazapyr, Penoxsulam, Triclopyr, 2,4-D, and Florpyrauxifen-benzyl are classified as systemic herbicides. They are absorbed through the photosynthetically active tissues of the plant and subsequently translocated to the site of action. In contrast, Diquat functions as a contact herbicide, exerting rapid phytotoxic effects relative to systemic compounds (Villamagna & Murphy, 2010; Kumar & Yadav, 2020).

Experimental investigations have further evaluated the impacts of glyphosate (Roundup), glufosinate-ammonium (Basta), and bispyribac sodium (Nominee) on water hyacinth, as well as on the survival of five aquatic invertebrate taxa inhabiting the Anzali Wetland: Hemiptera, Amphipoda, Odonata, Ostracoda, and Daphnia. Results indicated that all tested herbicides were effective in suppressing water hyacinth growth. Notably, glyphosate (Roundup) induced a significant reduction in shoot biomass, achieving a 98% score on the EWRC scale (Mirzajani et al., 2022).

Fig. 1 A mat of water hyacinth

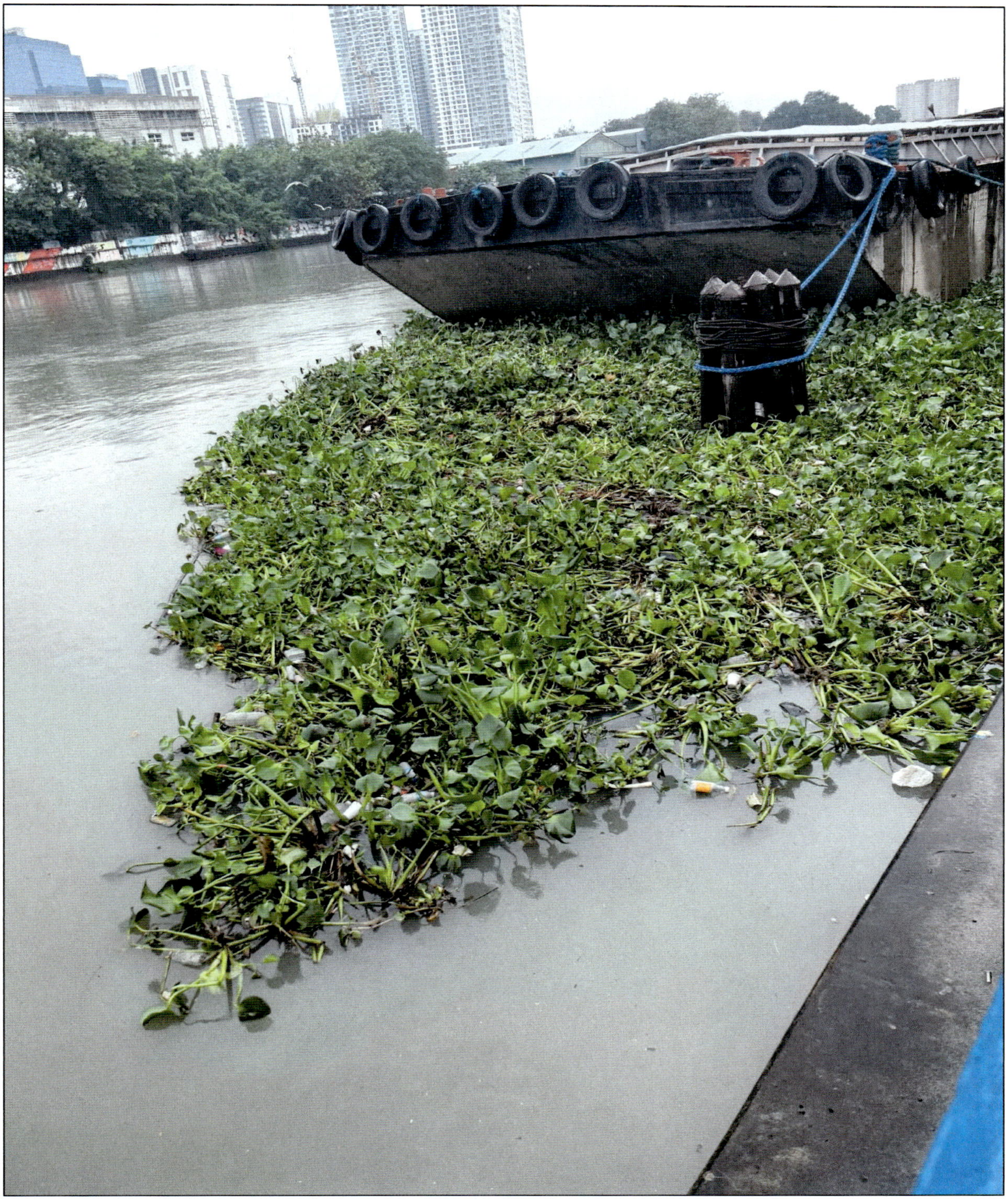

Fig. 2 This dense mat of water hyacinth encroaches upon a moored boat along a calm waterway.

Fig. 3 This water hyacinth reveals its spongy, bulbous, air-filled tissues, which provide buoyancy and enable the plant to float.

Invasive Species Concept Map

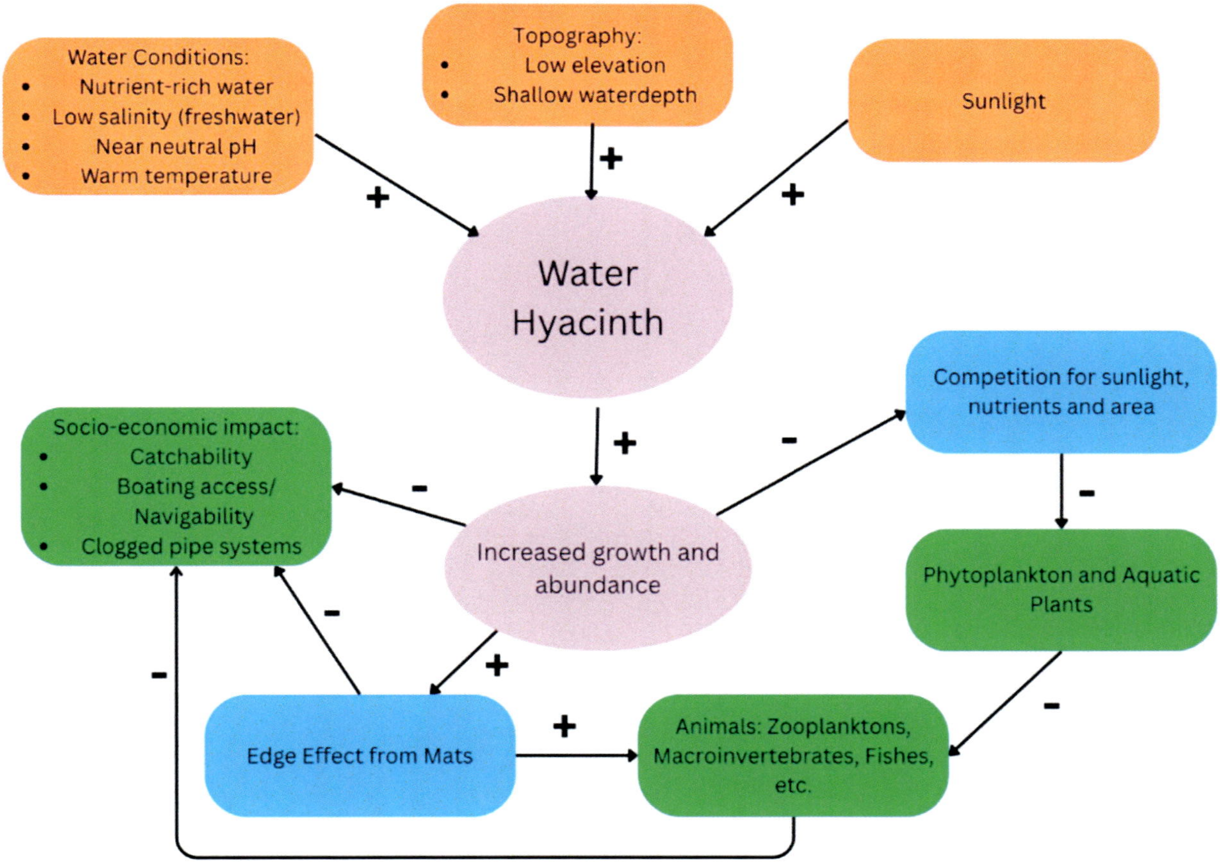

The illustration above shows Water hyacinth (*Pontederia crassipes* Mart.) thrives in freshwater environments characterized by low salinity, low elevation, shallow water depth, and warm tropical climates. Its proliferation is facilitated by favorable abiotic conditions such as abundant sunlight, nutrient availability, and being tolerant of polluted waterways. In the Bengawan Solo River in Indonesia, significant pollution was also correlated with the spread of *P. crassipes*, causing turbidity and pH disruption, a decrease in dissolved oxygen, and an increase in phosphate and nitrate levels (Islamy et al., 2024). It also proliferates under favorable topographic conditions, such as basin-shaped areas and floodplains, as observed in the Pasig River and Laguna de Bay. These factors collectively support rapid growth and reproduction, enabling the species to expand aggressively in terms of population size.

As a nutrient and light-demanding plant, water hyacinth competes directly with phytoplankton and native aquatic vegetation, thereby limiting their growth. Its dense floating mats provide temporary refuge for fish fry, invertebrates, and species such as Asian swamp eels, tilapia, catfish, whiskered terns, and little egrets. However, as coverage increases, primary producers decline, food webs are disrupted, and aquatic biodiversity is diminished. The mats also alter water quality by increasing turbidity, lowering dissolved oxygen, and modifying pH, nutrient concentrations, odor, and color. These changes degrade habitat quality and destabilize ecological relationships within lake and river systems (Wang & Yan, 2017).

The impacts extend beyond ecological communities to human populations. Dense infestations clog pipe systems, obstruct waterways, and hinder navigation, fishing, and irrigation (Villamagna & Murphy, 2010). Reduced fish catch and impaired access to aquatic resources highlight the socio-economic consequences of unchecked proliferation. In extreme cases, such as in flood-prone regions, the plant's dominance exacerbates the impacts of natural disasters by impeding water flow and intensifying damage.

Overall, water hyacinth exemplifies the paradox of invasive species: while it may provide limited ecological refuge, its dominance reshapes environmental conditions, undermines biodiversity, and disrupts both natural processes and human livelihoods. Its spread underscores the urgent need for integrated management strategies that address ecological, hydrological, and socio-economic dimensions of invasion.

References

Amédégnato, C., Devriese, H. (2008). Global diversity of true and pygmy grasshoppers (Acridomorpha, Orthoptera) in freshwater. FreshwaterAnimal Diversity Assessment, Springer Science & Business Media,Pp. 542. ISBN 978-1-4020-8259-7. doi:10.1007/s10750-007-9132-z.

Bashir, M.O.; El Abjar, Z. E. and Irving, N.S. (1984). *Observations on the effect of the weevils Neochetina eichhorniae Warner and Neochetina bruchi Hustache on the growth of water hyacinth. Hydrobiologia, 110*: 95-98, Limnology and Marine Biology in the Sudan.

Basilio, J. A. M., & Buot Jr, I. E. (2023). PONTEDERIA CRASSIPES MART.(PONTEDERIACEAE): ITS BIOLOGY AND SOCIO-ECOLOGICAL IMPACTS IN THE PHILIPPINES.

Barrett, SCH. (1980). *Sexual reproduction in Eichhornia crassipes (water hyacinth). II. Seed production in natural populations. J Appl Ecol.* [accessed 2022 Dec 13];17:113124. doi:10.2307/2402967.

Calonzo, A. (2011). River-clogging plant causes devastating Cotabato flood. GMANewsOnline. https://www.gmanetwork.com/news/topstories/regions/223609/r iver-clogging-plant-causes-devastating-cotabato-flood/story/

Charudattan, R., Labrada, R., Center, T. D., & Kelly-Begazo, C. (Eds.). (1996). Strategies for water hyacinth control: Report of a panel of experts meeting, 11–14 September 1995, Fort Lauderdale, Florida, USA. Food and Agriculture Organization of the United Nations. https://www.fao.org/fileadmin/templates/agphom e/documents/Biodiversity-pollination/Weeds/Docs/Water_Hyacinth.pdf

Coetzce, J.A., Byrne, M.J., Hill, M.P. (2007). *Predicting the distribution of Eccritotarsus catarinensis, a natural enemy released on water hyacinth in South Africa. Entomol, Exp.Appl 125*(3), 237-247

Day, M. D., den Breeyen, A., Joshi, R. C., & Dela Cruz, M. S. (2023). *First report of establishment of two weevils, Neochetina bruchi Hustache and Neochetina eichhorniae Warner (Coleoptera: Curculionidae), released against water hyacinth (Pontederia crassipes Mart.) in the Philippines. Weeds – Journal of the Asian-Pacific Weed Science Society, 5*(2), 21–26. https://era.daf.qld.gov.au/id/eprint/12810/1/Day_etal%20Hyacinth_Neochetina%2 02023.pdf

Dersseh, M. G., Melesse, A. M., Tilahun, S. A., Abate, M., & Dagnew, D. C. (2019). Water hyacinth: review of its impacts on hydrology and ecosystem services—lessons for management of Lake Tana. Extreme hydrology and climate variability, 237-251.

Eppo.int. (2019). EPPO Global Database. [online] Available at: https://gd.eppo.int.

Feedipedia. (n.d.). Water hyacinth (Eichhornia crassipes). https://www.feedipedia.org/node/160

Francisco, MA. (2020). Why is the water hyacinth a notorious aquatic pest?. Pasig City (Philippines): FlipScience. [accessed 2023 Jun 23]. https://www.flipscience.ph/ plants-and-animals/water-hyacinth/.

Gupta, A. K., & Yadav, D. (2020). *Biological control of water hyacinth. Environmental Contaminants Reviews, 3*(1), 37–39. https://doi.org/10.26480/ecr.01.2020.37.39

Harun, I., Pushiri, H., Amirul-Aiman, A. J., & Zulkeflee, Z. (2021). *Invasive water hyacinth: Ecology, impacts and prospects for the rural economy. Plants, 10*(8), 1–23.

Hasan, M. R., & Chakrabarti, R. (2009). *Use of algae and aquatic macrophytes as feed in small-scale aquaculture-A review.*

Hill, M., Coetzce, J. (2008). Integrated control of water hyacinth in Africa. EPPO Bull 38 (3), 452-457

Islamy, R. A., Hasan, V., Kilawati, Y., Maimunah, Y., Mamat, N., & Kamarudin, A. S. (2024). *Water hyacinth (Pontederia crassipes) bloom in Bengawan Solo River, Indonesia: An aquatic physicochemical and biology perspective. International Journal of Conservation Science, 15*(4), 1885-1898.

Jacinto & Lirio. (n.d.). Environmental impact of water hyacinths Philippines. https://jacintoandlirio.com/pages/environmental-impact

Jha, K., Wu, B., Aziz, S. Q., Ahmed, H. G., & Ahmed, M. S. (2024). *Sustainable solutions for water hyacinth invasion: Characteristics, impacts, control, and utilization. World Journal of Advanced Engineering Technology and Sciences, 12*(1), 47–58. https://doi.org/10.30574/wjaets.2024.12.1.0169

Julien, M. H. (2001). Biological control of water hyacinth with arthropods: A review to 2000. Canberra: Australian Centre for International Agricultural Research (ACIAR). Retrieved from http://www.bio-nica.info/Biblioteca/Julien2001Eischhornia.pdf

Khan, H., Uslu, Ö. S., & Gul, B. (2022). Management and utilization of water hyacinth (Eichhornia crassipes Mart.). In Ö. Toncer & Z. Gunes (Eds.), A look into some recent advances in biology, ecology and agricultural practices (Chapter 4).

IKSAD Publishing House. https://www.researchgate.net/publication/366593455 Living Lakes Network. (n.d.). Laguna de Bay. Living Lakes. https://livinglakes.org/laguna-de-bay/

Lu, X., Kruatrachue, M., Pokethitiyook, P., & Homyok, K. (2004). Removal of cadmium and zinc by water hyacinth (Eichhornia crassipes). ScienceAsia, 30(2), 93–103. https://doi.org/10.2306/scienceasia1513-1874.2004.30.093

Manila Bulletin. (2023). Not out of this world: The danger posed by invasive species. https://mb.com.ph/2023/8/11/not-out-of-this-world-the-danger-posed-by-invasive-species Mandaluyong City Government. (n.d.). Biological environments. https://mandaluyong.gov.ph/profile/biological-environments

Mirzajani, A., Nezamabadi, N., Salavatian, S., Bagheri, S., Salahi, M., & Rufchaei, R. (2022). Chemical control of water hyacinth by some herbicides and their effect on some aquatic invertebrate.

Pieterse, A. H. (1997). Eichhornia crassipes (Martius) Solms. Record from Proseabase. Faridah Hanum, I & van der Maesen, L.J.G. (Editors). PROSEA (Plant Resources of South-East Asia) Foundation, Bogor, Indonesia

Philstar Interaksyon. (2019). Freshwater fish tilapia caught in Pasig River after years of rehabilitation. https://interaksyon.philstar.com/trends-spotlights/2019/01/29/143140/freshwater-f ish-tilapia-caught-in-pasig-river-after-years-of-rehabilitation/

Stiefel, K. (2023). How did the janitor fish end up surviving—and thriving—in the Pasig River? Esquire Philippines. https://www.esquiremag.ph/long-reads/features/janitor-fish-pasig-river-a00048-20 230607-lfrm

U.S. Fish & Wildlife Service. (2020). Water hyacinth acts like 'plastic wrap' on the Delta. https://www.fws.gov/story/2020-07/water-hyacinth-acts-plastic-wrap-delta

Wang, Z., & Yan, S. H. (2017). Direct and strong influence of water hyacinth on aquatic communities in natural waters. In Water Hyacinth (pp. 44-65). CRC Press.

Wimalarathne, H., & Perera, P. (2019). *Potentials of water hyacinth as livestock feed in Sri Lanka. Indian Journal of Weed Science, 51*(2), 101–105.

Velasquez, A., Mercado, V., Marquez, C., & Moreno, M. L. (2024). Species abundance and diversity of avifauna in the Pasig River, Philippines. Paper presented at the 10th Asian Wetlands Symposium, Crown Plaza Manila Galleria, Philippines.

Villamagna, A. M., & Murphy, B. R. (2010). *Ecological and socio-economic impacts of invasive water hyacinth (Eichhornia crassipes): a review. Freshwater biology, 55*(2).

Yang, H., Fang, X., Chen, Z. (2001). Integrated farming systems. In: Integrated agriculture-aquaculture: a primer. FAO Fisheries Technical Paper, No. 407. FAO, Rome, Italy

Thunbergia grandiflora Roxb. A Beautiful Disaster

Jayvee Velasco Mendoza

Species Name

Thunbergia grandiflora Roxb. is most known as "Sky Flower" but is also called Blue Trumpet Vine due to its bluish flowers assuming the shape of a trumpet, Clock Vine, Bengal Clock Vine, and Sky Vine, among others (Missouri Botanical Garden, n.d.; National Parks, n.d.). The species was named by William Roxburgh in 1820 and was published in Edward's Botanical Register (International Plant Names Index, n.d.).

Species Distribution

According to Royal Botanic Gardens Kew (n.d.), *Thunbergia grandiflora* is "native" to Assam, Bangladesh, Cambodia, China South-Central, China Southeast, East Himalaya, Hainan, Laos, Malaya, Myanmar, Nepal, Thailand, and Vietnam. Meanwhile, the species is "introduced" in the countries of Ascension, Belize, Cameroon, Caroline Is., Comoros, Cook Is., Costa Rica, Cuba, Dominican Republic, Ecuador, El Salvador, Fiji, Florida, Gabon, Gambia, Guatemala, Gulf States, Haiti, Hawaii, India, Jamaica, Leeward Is., Madeira, Marianas, Mauritius, Panamá, Paraguay, Puerto Rico, Réunion, Society Is., Sri Lanka, Trinidad-Tobago, Venezuelan Antilles, and Windward Island. Furthermore, according to Co's Digital Flora of the Philippines (n.d.), the species is "naturalized" in the Philippines.

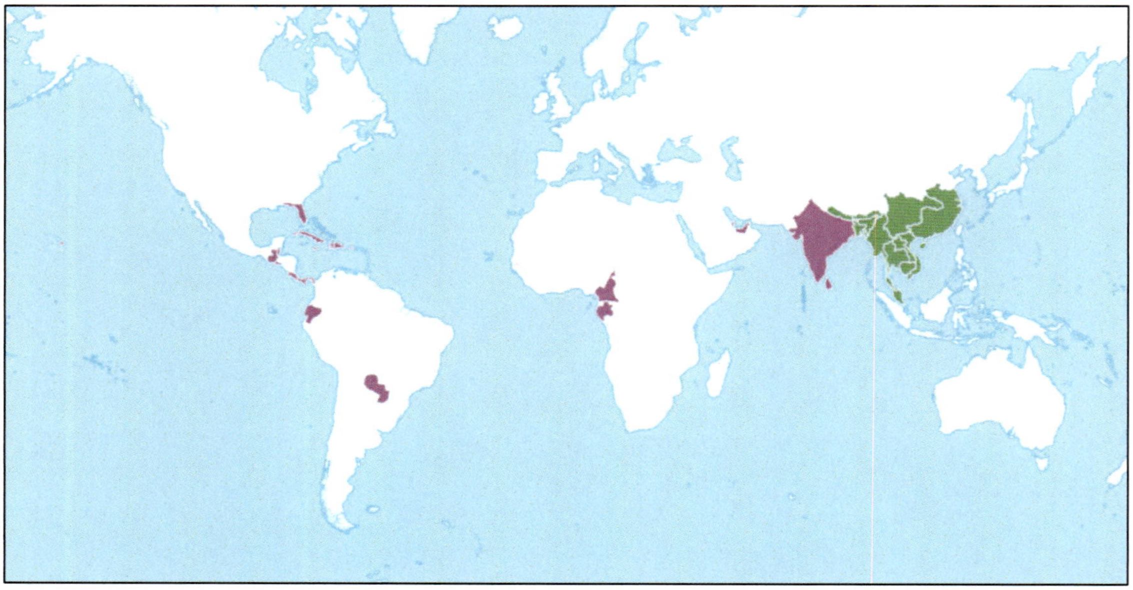

Fig. 1 The global distribution map of *Thunbergia grandiflora* Roxb. (Royal Botanic Gardens Kew, n.d.).

On the other hand, the species is utilized as an ornamental plant, which refers to those plants that are cultivated primarily for their aesthetic value that comes from its scent, color and are used as decorations in households both indoors and outdoors (Dauda et al., 2023), which therefore serves as its primary method of introduction (Sandoval and Rodriguez, 2012). Moreover, among the species' earliest records of introduction occurred during the late 19th century in the Caribbean Region and were accounted as "cultivated in gardens and escaped" by Ignaz Urban in his book "Symbolae Antillinae," specifically for the islands of Cuba, Guadeloupe, Dominica, Martinique, and St. Vincent (Sandoval and Rodriguez, 2012). The species' introduction was also accounted for in Bermuda, Puerto Rico, the Virgin Islands, and Hawaii before the 1990s (Sandoval and Rodriguez, 2012). Based on this, Sandoval and Rodriguez (2012) inferred that the species' invasiveness, given the ease of its introduction and its high dispersal ability, is high.

Ecological Impacts

Thunbergia grandiflora reproduces sexually through seeds and by cuttings, both of which contribute to its rapid growth and invasion in the areas where it develops (Sandoval and Rodriguez, 2012). From this point, the species continuously invades and outcompetes native plants and vegetation in the area, obliterating host trees and understory species and eventually inhibiting the germination and growth of the native species' seedlings it competes with (Sandoval and Rodriguez, 2012).

Furthermore, the species is found to have the ability to completely invade acres of vegetation, like it did in some areas of Australia, and does not support any native wildlife, which makes it a much more serious ecological threat, especially to native species in lowland tropical rainforests (Natural Resources Conservation Service, n.d.; Sandoval and Rodriguez, 2012).

Economic Impacts

The species is used as an ornamental plant, making it a good product for small businesses and entrepreneurs. Furthermore, the species is also found to be medicinally useful and as an organic fertilizer (BioNET-EAFRINET, 2011).

Control and Eradication Measures

Physical/Mechanical Control

The species can be eradicated by manually removing small host plants, including their roots and all plant segments, to prevent resprouting (Sandoval and Rodriguez, 2012). For larger host plants or trees, specialized machinery and equipment are deemed necessary, as manual removal will be difficult (Sandoval and Rodriguez, 2012). Lastly, the species should not be promoted for ornamental use at all in areas or regions that are within or near natural forests where several native species are thriving (Sandoval and Rodriguez, 2012).

Chemical Control

The use of herbicides such as glyphosate has been shown to be an effective chemical control against the species (Sandoval and Rodriguez, 2012). Imazapyr, another chemical herbicide, has also been found to be effective in eradicating the species while being gentle towards vegetation and other plant species within or near it (Sandoval and Rodriguez, 2012).

Fig. 2 *Thunbergia grandiflora* Roxb.

Concept Map

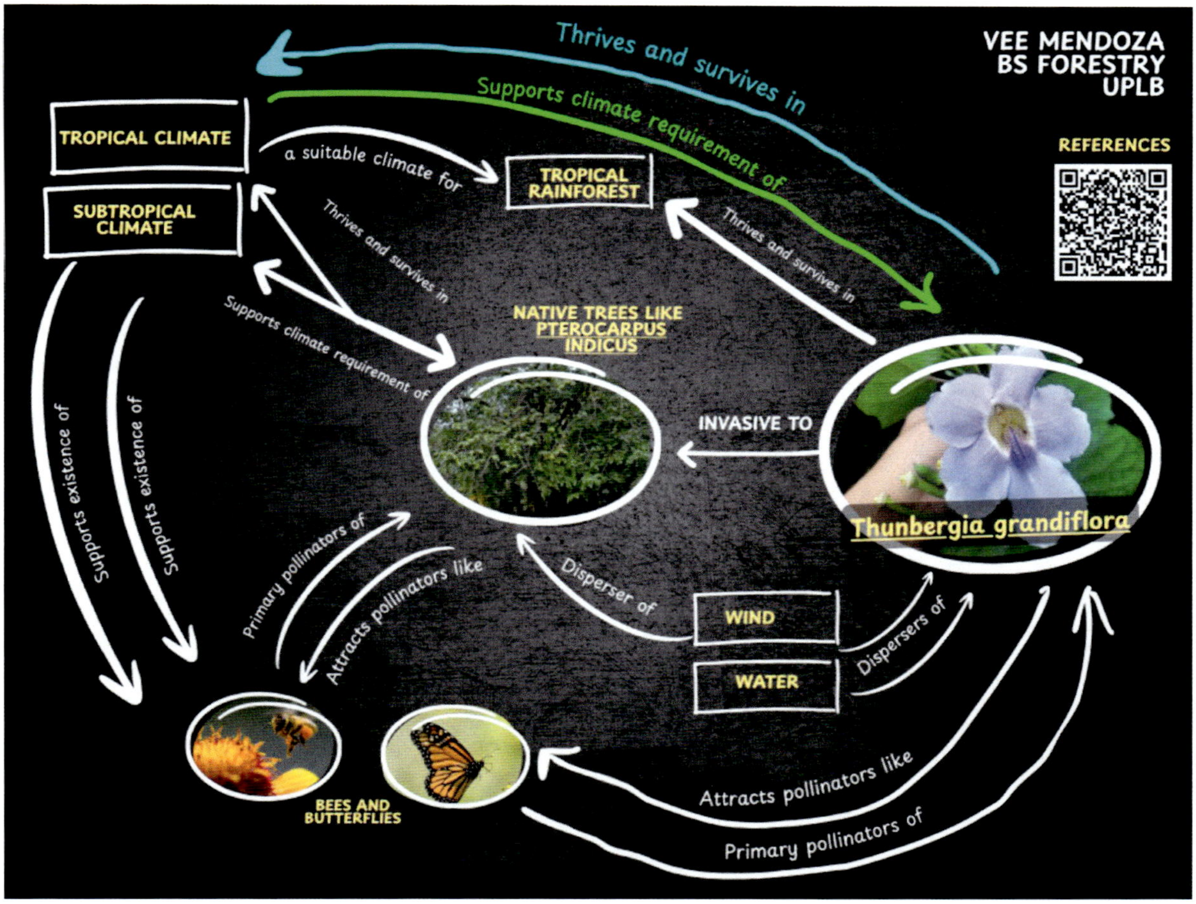

The concept map above aims to explain the ecological relationships of *Thunbergia grandiflora* with biotic and abiotic factors, including insects, other plant species, climate, and other environmental elements.

It is evident from the map that *T. grandiflora* is an invasive relative of *Pterocarpus indicus* because it also thrives under the climate conditions and forest types where the latter species is thriving. Additionally, the pollinators and dispersers of the two species are also similar, which therefore makes the *T. grandiflora* a threat to the ecological stability of the native species, like *P. indicus*, with which it competes.

References

BioNET-EAFRINET (The East African Network for Taxonomy), 2011. Thunbergia grandiflora (Blue Thunbergia). Keys and Fact Sheets for Invasive Plants. http://keys.lucidcentral.org/keys/v3/eafrinet/weeds/key/weeds/Media/Html/Thunbergia_grandiflora_%28Blue_Thunbergia%29.htm

Co's Digital Flora of the Philippines. (n.d.). Acanthaceae. https://www.philippineplants.org/Families/Acanthaceae.html

Dauda, W. P., Dantata, I. J., Adetunji, C. O., Abraham, P., Ifeanyi, U. J., Glen, E., Morumda, D., Abraham, S. E., Wabba, G. P., Ogwuche, I. O., & Ogundolie, F. A. (2023). The effect of evaporative coolant structure on ornamental plants. In Elsevier eBooks (pp. 209–235). https://doi.org/10.1016/b978-0-323-89864-5.00011-4

International Plant Names Index. (n.d.). Thunbergia grandiflora. https://www.ipni.org/n/117460-3 Natural Resources Conservation Service. (n.d.). Blue Trumpet Vine (Thunbergia grandiflora). Invasive Species Fact Sheet Pacific Islands Area. https://www.hicattle.org/Media/HICattle/Docs/thurnbergia_grandiflora.pdf

Royal Botanic Gardens Kew. (n.d). Thunbergia grandiflora Roxb. Plants of the World Online. https://powo.science.kew.org/taxon/urn:lsid:ipni.org:names:117460-3#publications

Sandoval, J., & Rodríguez, P. (2012). Thunbergia grandiflora (Bengal trumpet) [Dataset]. https://doi.org/10.1079/cabicompendium.117524

From History to Management: Understanding the Invasiveness and Impacts of *Mikania micrantha* Kunth on Philippine Biodiversity and Crop Productivity

Dailyn Rui C. Aguila & Russell Ivan D. Bautista

Introduction: Species History, Information, and Characteristics

Mikania micrantha is a fast-growing, aggressive climber. In the Philippines, it is commonly called bikas (Bagobo), detidid (Igorot), and uoko (Bontoc), while common English names include bitter vine, mile-a-minute weed, plant-killer, Chinese creeper, climbing hempvine, and American rope.

Fig. 1 *Mikania micrantha*

Native to South America and the Caribbean, *M. micrantha* is an invasive vine that has become a significant ecological concern in many tropical regions, including the Philippines. *M. micrantha*'s remarkable biological characteristics include rapid, highly successful sexual and vegetative production, and broad environmental tolerance, enabling it to spread quickly and dominate landscapes, effectively displacing native species. These traits enable the species to form dense mats that cover and smother other plants, thereby preventing them from accessing sunlight and essential nutrients. As a result, this species profoundly alters ecosystem balance and interacts with various traditional, economic, ecological, environmental, and biological components.

Traditionally, it has been used across regions to treat various ailments. Ethnobotanical practices document their use in managing insect bites, skin rashes, itching, chickenpox, wounds, sores, fever, stomachache, colds, jaundice, rheumatism, and respiratory diseases (Ali Khan et al., 2023). In

Jamaica, the leaf juice is applied for wound dressing and healing of sores, while in Assam (Northeast India), the Kabi tribes use it as an antidote for insect bites, scorpion stings, and minor bleeding (Ali Khan et al., 2023). Additional medicinal applications reported in Malaysia include the management of cholesterol, high blood pressure, high blood glucose levels, and general itching. Across Africa, the leaves are consumed as a green vegetable, and in Malaysia, the species is planted as an understory cover in rubber plantations and on slopes to reduce soil erosion (Ali Khan et al., 2023; Poudel et al., 2019).

Economically, the benefits derived from *M. micrantha* are limited compared to its ecological impacts. It is occasionally used as fodder for sheep and cattle, particularly in Malaysia and parts of India, as an alternative during periods of grass scarcity. It can serve as green manure or organic fertilizer to improve crop yields (Poudel et al., 2019). Despite these uses, the species is known to cause hepatotoxicity in dairy cattle, and its rapid growth contributes to extensive infestations in forests, plantations, and disturbed habitats. The plant is also susceptible to infection by *Mikania micrantha* wilt virus (MMWV), a Fabavirus that may affect its physiological performance (Ali Khan et al., 2023). Overall, *M. micrantha* demonstrates both ecological invasiveness and notable pharmacological promise, positioning it as a species of dual significance that requires integrated management strategies alongside continued scientific investigation.

From an ecological standpoint, the invasion of *M. micrantha* produces both positive and negative effects. On the one hand, the species can provide a temporary food source for certain insects and a short-term habitat for small animals, indicating its limited contribution to certain ecological functions. However, its negative impacts are far more extensive. The vine competes aggressively with native plants, reducing biodiversity and altering community structures. Its dominance can also lead to increased soil erosion once native vegetation dies, as root systems that stabilize the soil are lost. These changes disrupt ecological stability and reduce the ability of natural systems to recover from disturbances.

In terms of environmental impacts, *M. micrantha* can slightly improve soil fertility in the short term by decomposing organic matter, which adds nutrients to the soil. Its flowers may also attract some pollinators. However, the adverse effects are far more severe. The plant alters soil composition and microbial balance, affecting nutrient cycling and soil health. It also disrupts pollination networks by outcompeting native flowering plants that pollinators depend on. Furthermore, by smothering native vegetation, *M. micrantha* decreases the diversity and productivity of natural habitats, undermining ecosystem resilience.

The biological impacts of *M. micrantha* are similarly complex. While it can serve as a food source for certain insects and decomposers, it negatively affects overall faunal diversity in ecosystems. The decline in habitat quality and the loss of native flora reduce food availability for herbivores and other dependent species. Human populations are not spared either; contact with the plant may cause skin irritation or allergies, and communities that depend on forest and agricultural resources face reduced crop yields and higher weed management costs. These effects illustrate the intertwined relationship between ecological health and human well-being.

Finally, *Mikania micrantha*'s relationship with other organisms demonstrates how one species can significantly reshape entire ecosystems. Although it provides shelter for some small animals and supplies organic carbon to soil microorganisms, it also displaces native flora, which are vital to the survival of many species. The overall impact of this species is thus characterized by ecosystem imbalance, reduced biodiversity, and disrupted ecological interactions. Its presence highlights the delicate interconnectedness of living organisms within ecosystems, showing how the dominance of one invasive species can cascade across biological, environmental, and human dimensions, ultimately threatening the stability and sustainability of natural systems.

Species' Invasiveness

This creeping herb is known for its rapid invasion, colonizing areas in both terrestrial and wetland ecosystems. It was observed that *M. micrantha* highly favors flowering in open sunny areas rather than in shaded areas (Hong et al., 2010; Alcalde & Buot, 2023), making it a successful colonizer.

Mikania micrantha reproduces both vegetatively via stem fragments and sexually via seeds. Its flowers are arranged in a corymb and typically bloom during the dry season. The species also possesses a pappus, a highly modified calyx of the daisy family, which, in *M. micrantha*, is terminal (Yang et al., 2005; Alcalde & Buot, 2023). As a result, these structures enable their seeds to be dispersed by the wind. These pappus, according to Yang et al. (2005), were also reported to attach to human clothing and equipment, as well as to animal fur, thereby making humans and animals their agents of dispersal. On the other hand, stem nodes, when in contact with the soil's surface, lead to resprouting of the vine (Day et al., 2012; Alcalde & Buot, 2023). The main stem, according to Puzari et al. (2010) as cited by Alcalde and Buot (2023), can withstand drought for years before reshooting under suitable conditions.

Native Range and Method of Introduction

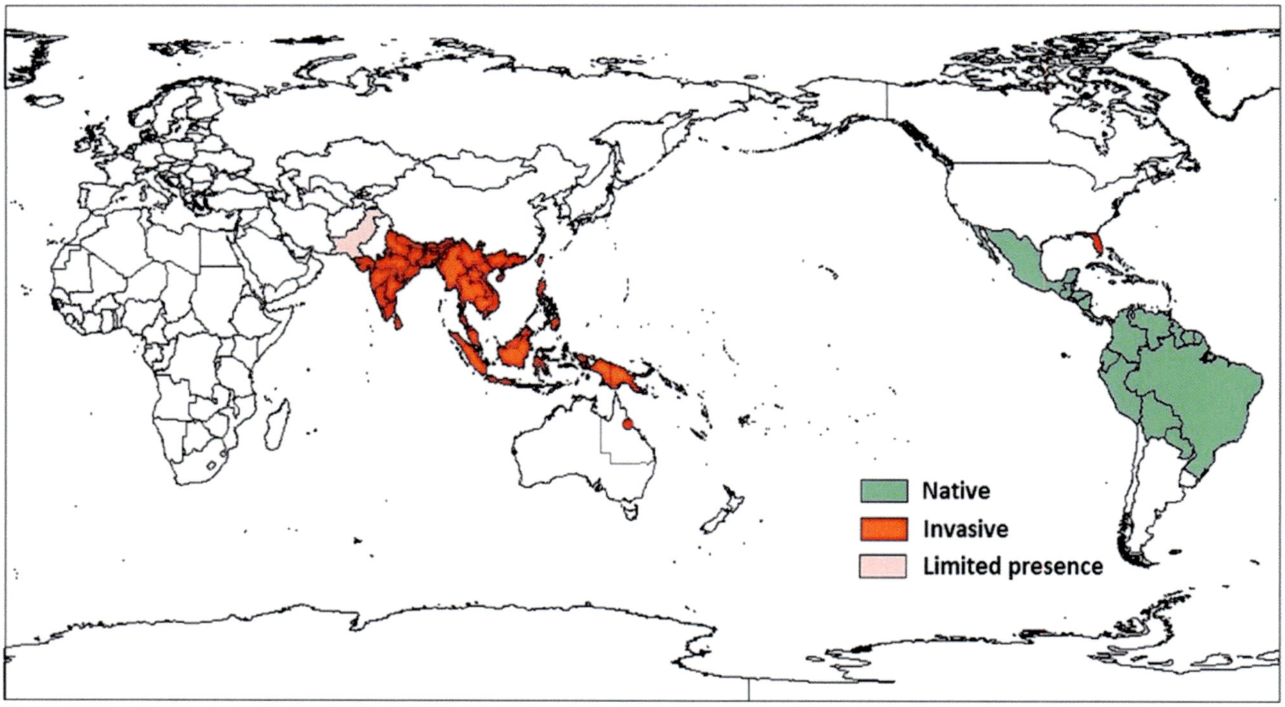

Fig. 2 Native and introduced habitat ranges for *M. micrantha*, adapted from Day et al. (2016)

According to Plants of the World Online (POWO), the native range of this species is Tropical & Subtropical America. *Mikania micrantha* is native to a broad geographic range spanning Central America, South America, parts of the Caribbean, and the southeastern United States. Its native distribution includes: Argentina (Northeast and Northwest), Belize, Bolivia, Brazil (North, Northeast, South, Southeast, and West-Central), Colombia, Costa Rica, Cuba, the Dominican Republic, Ecuador, El Salvador, Florida, French Guiana, Guatemala, Guyana, Haiti, Honduras, Jamaica, the Leeward Islands, Mexico (Central, Gulf, Northeast, Northwest, Southeast, and Southwest), Nicaragua, Panama, Paraguay, Peru, Puerto Rico, Suriname, Trinidad and Tobago, Uruguay, Venezuela, and the Windward Islands. This extensive native range reflects the species' adaptability to diverse tropical and subtropical environments.

From this broad native distribution, some anecdotes suggest it was intentionally introduced into numerous countries, while it was accidentally introduced into others. The species is now established in Assam, the Caroline Islands, the Chagos Archipelago, southeastern China, Christmas Island, the eastern Himalayas, Fiji, India, the Laccadive Islands, Malaya, Mauritius, Myanmar, the Nansei-shoto Islands, Nepal, Niue, the Philippines, Queensland (Australia), Réunion, Samoa, Sumatra, Taiwan, Thailand, Tokelau-Manihiki, Vanuatu, and the western Himalayas.

In a study by Day et al. (2016), as cited by Banerjee et al. (2019), the invasiveness of *M. micrantha*, although speculative and based on anecdotal evidence, suggests that multiple introductions were geographically and temporally separated. It was introduced primarily for horticultural use, soil conservation initiatives, accidental transport, and the movement of goods and equipment, all of which facilitated its rapid naturalization. Once established, the vine expanded aggressively due to its rapid growth, efficient vegetative reproduction, and the absence of natural enemies in the invaded landscapes, making physical control and eradication nearly impossible.

In the Philippines, *M. micrantha* was intentionally introduced, likely as a cover crop or ornamental species for gardens. Based on an analysis of archival literature and herbarium catalogs, *M. micrantha* was first recorded in the Philippines, particularly in Luzon province, in 1838 (Banerjee et al., 2019). This was the first established locality outside its native range in Central and South America. Its introduction likely occurred via the Pacific trade routes linking European countries and Latin America, with the Philippines, through its Manila Galleon trade, serving as an important trading post (Legarda, 1999; Banerjee et al., 2019). From the Philippines, the species began spreading to the Malay Peninsula, as evidenced by the influx of floristic records from that period in the region.

Ecological and/or Economic Impacts

According to Zhang et al. (2004), *Mikania micrantha* is widely regarded as one of the most destructive invasive weeds globally (Holm et al., 1977). Its rapid growth rate, prolific reproduction, and aggressive smothering behavior enable it to dominate and eliminate native vegetation by obstructing sunlight (Huang et al., 2000). These characteristics mirror the well-documented capacity of alien plant species to alter community structure and ecosystem functioning (Higgins et al., 1996).

Ecological Impacts

The vine invades a broad range of habitats, including orchards, forest edges, riverbanks, disturbed sites, and roadsides (Kong et al., 2000a). It is also observed in fence lines and extends into aquatic habitats, such as ponds, where it covers and kills aquatic life (Zhang et al., 2004).

In a study cited by Alcalde and Buot (2023), the species can also modify soil chemistry and mineral cycling, thereby disrupting microbial populations in soil (Wong, 1964). Similarly, it inhibits the growth of nearby crops, thereby severely affecting crop yields and, in some cases, killing them (Day et al., 2012).

Economic Impacts

Economically, the species is a significant pest in plantation agriculture, particularly in tea, teak, rubber, and oil palm systems (Waterhouse, 1994). Generally, it competes directly with crops, reduces yields, interferes with management operations, and increases labor costs.

In regions such as northern Australia, its potential introduction poses a major threat to high-value ecosystems, including the Queensland wet tropics (Queensland Government Natural Resources and Mines, 2003).

In another study (Widjaja & Tjitrosoedirdjo, 1991) focusing on the effects of *M. micrantha* in a bamboo plantation, the said species has covered the plantation despite the competition with other present weeds, hence it has led to the stunted growth of the bamboos, and it has disrupted their development (Alcalde & Buot, 2023). In China, Shen et al. (2013) reported that *M. micrantha* had invaded agricultural crops, including sugarcane, bamboo, lemon, banana, and orange, leading to reduced crop yields. In Malaysia, the girth of rubber trees in plantations with *M. micrantha* was found to be 27% thinner than that of trees in plantations with leguminous ground covers (Watson et al., 1964; Alcalde & Buot, 2023).

In the Philippines, the economic impact of *M. micrantha*'s invasiveness remains understudied, as Alcalde and Buot (2023) revealed that the species receives limited attention from the government and research community. However, it is generally true that the economic impacts of invasive species have been studied only to a limited extent in the country. Various studies from neighboring countries have examined the effects of *M. micrantha*'s invasiveness on agricultural commodities. It is important to note that these crops are also economically important in the Philippines, and controlling such species, regarded as invasive in different ecosystem habitats, would be costly.

Restoring Balance: Control and Eradication Measures

Various control and eradication measures were developed to address the invasive potential of *M. micrantha*, including mechanical, chemical, and biological control. However, due to its strong sexual and asexual reproduction, morphological plasticity, and adaptive evolution, the use of a single control measure cannot effectively eradicate the damage caused by *M. micrantha* and hence requires adopting more effective and comprehensive prevention and control measures (Shen et al., 2013; Bora et al., 2023).

Mechanical Control

For mechanical eradication of *M. micrantha*, hand-pulling, digging, uprooting, and slashing were used. Manual removal remains the most widely practiced method of managing *Mikania micrantha*. The methods discussed above, however, were very labor-intensive and uneconomical, as they required regular maintenance and repetitive processing. Specifically, manual cutting of vines close to the ground is recommended monthly for three consecutive months during summer and autumn, followed by continued removal in winter and spring. This approach is favored over burning or machine-based clearing, as the species can readily regenerate from deep root systems and stem fragments (Swamy & Ramakrishnan, 1987b; Zhang et al., 2004). The use of manual and mechanical control for this species also became ineffective due to rapid regrowth from cut stumps, as well as accidental seed dispersal caused by physical and mechanical disturbances.

Biological Control

Some biological control strategies for eradicating *M. micrantha* include fungi, certain insects, and other tall grasses that could outcompete it. Some attempts at experimentation with insects as biocontrol agents have failed.

One of the most documented biological controls and sustainable long-term strategies for managing *M. micrantha* is particularly important, as the species arrived in China without its native natural enemies (Tu et al., 2001). Several insect and fungal organisms have been observed feeding on or damaging the species, including:

- *Aphis citricola* – causes leaf curling (Han et al., 2001)
- *Pseudocerospora sp.* – damages young leaves, especially under moist conditions
- *Brevipalpus phoenicis* – induces leaf scorching (Chen et al., 2002; Zhou et al., 2003)
- *Acanthopsyche sp.* – feeds on plant tissue (Shao et al., 2002)

Classical biological control efforts have identified several promising agents from the species' native range. *Liothrips mikaniae*, a natural enemy identified by Cock (1982), has shown potential but failed to establish following releases in the Solomon Islands and Malaysia.

Two nymphalid butterflies, *Actinote anteas* and *Actinote thalia pyrrha*, known to feed on *M. micrantha* in Indonesia, are subjected to rearing and evaluation (Chenon, 2002; Li et al., 2002; Shao et al., 2002b) (Zhang et al., 2004).

Other countries have reported a highly damaging microcyclic rust, *Puccinia spegazzinii*, as a primary biocontrol agent for *M. micrantha* due to its high virulence, particularly in the Neotropics, where it is undergoing host-specificity testing (Barreto & Evans, 1995; Cock et al., 2000). However, there are no records yet that the fungus has been used in the Philippines. Bora et al. (2023) reported that *P. spegazzinii* is naturally occurring in the neotropics and damages *M. micrantha* by causing necrosis and canker in all aerial parts of the plant, ultimately killing it.

On the other hand, tall grasses having long, elastic leaf blades or stalks were usually the natural biocontrol agents in eradicating *M. micrantha*. With these features, the vine's growth and suppression would be limited, as it could not grasp and climb, thereby largely reducing the species' biomass. Some of these grasses include *Megathyrsus maximus* (formerly *Panicum maximum*) and *Cenchrus purpureus* (formerly *Pennisetum purpureum*), both of which were recorded as naturalized species in the Philippines according to the Co's Digital Flora of the Philippines, a free, accessible, and illustrated database of Philippine vascular plants.

Chemical Control

Chemical methods can suppress dense infestations, although they rarely achieve complete eradication. Herbicides such as glyphosate, 2,4-D, napropamide, and atrazine are effective to varying degrees (Palit, 1981). Experimental evaluations show that glyphosate, bentazone, picloram, and oxadiazon inhibit seed germination, with 0.4% bentazone and 0.2% picloram performing best in laboratory and field trials (Hu & But, 1994).

Field studies in Shenzhen demonstrated that while 2,4-D and glyphosate killed above-ground biomass, plants recovered within three months. By contrast, 25% hexazinone (injected) and 75% sulfometuron methyl (spray-treated) effectively killed entire plants within 2–5 months (Zan et al., 2001; Zhang et al., 2004).

Allelopathic suppression is also a potential chemical pathway. *Delonix regia* leaf and flower powders resulted in 75–90% seedling mortality, while its aqueous extracts produced high phytotoxicity, indicating potential value for integrated weed management (Kuo et al., 2002) (Zhang et al., 2004).

Usual herbicides were used as chemical control for *M. micrantha*. The application of herbicides to control *M. micrantha* is preferred over other control methods because it offers a quicker, more effective, and relatively cheaper means of control (Zhang et al., 2004; Bora et al., 2023). In the Philippines, herbicides such as glyphosate and 2,4-D are used; however, their use and recommendations are regulated by the Philippine Fertilizer and Pesticide Authority (FPA).

Since its introduction to a wide range of areas, *Mikania micrantha* has undergone extensive spatial expansion supported by its biological characteristics, the absence of natural enemies, and increased anthropogenic disturbance. Its capacity to dominate a wide array of habitats and negatively affect ecosystems and agricultural systems makes it a serious invasive threat. While eradication remains challenging, integrated strategies that combine manual removal, habitat modification, targeted chemical applications, and ongoing research on biological control agents offer the most viable framework for long-term management.

Visual Highlights of Field Observations: Original Field Photos

The field observations were conducted at two sites: the University of the Philippines Los Baños (UPLB), Laguna, and Tagaytay City, Cavite. Both localities were in Region IV-A (CALABARZON), in the island of Luzon, the Philippines.

Fig. 3 a) adaxial leaf surface of *Mikania micrantha*, b) abaxial leaf surface of *M. micrantha*, c) stem of *M. micrantha* with the presence of trichomes, d) *M. micrantha* showing its twining stems

Fig. 4 *Mikania micrantha* in its habitat at the back of the main library on the University of the Philippines Los Baños campus

Fig. 5 Invasiveness of *Mikania micrantha* in Tagaytay City, Cavite, Philippines: (A) *M. micrantha* showing its invasive capability by slowly invading an abandoned residential house, (B) *M. micrantha* creeping in a rusted window

Fig. 6 Close-up view of *Mikania micrantha* leaves: (A) *M. micrantha* leaves with its creeping habit, (B) Leaves of *M. micrantha* in a rusted window

Fig. 7 Close-up view of the inflorescences of *Mikania micrantha*

Overview of Invasive Species' Key Concepts and Relationships

Invasive Species Concept Maps

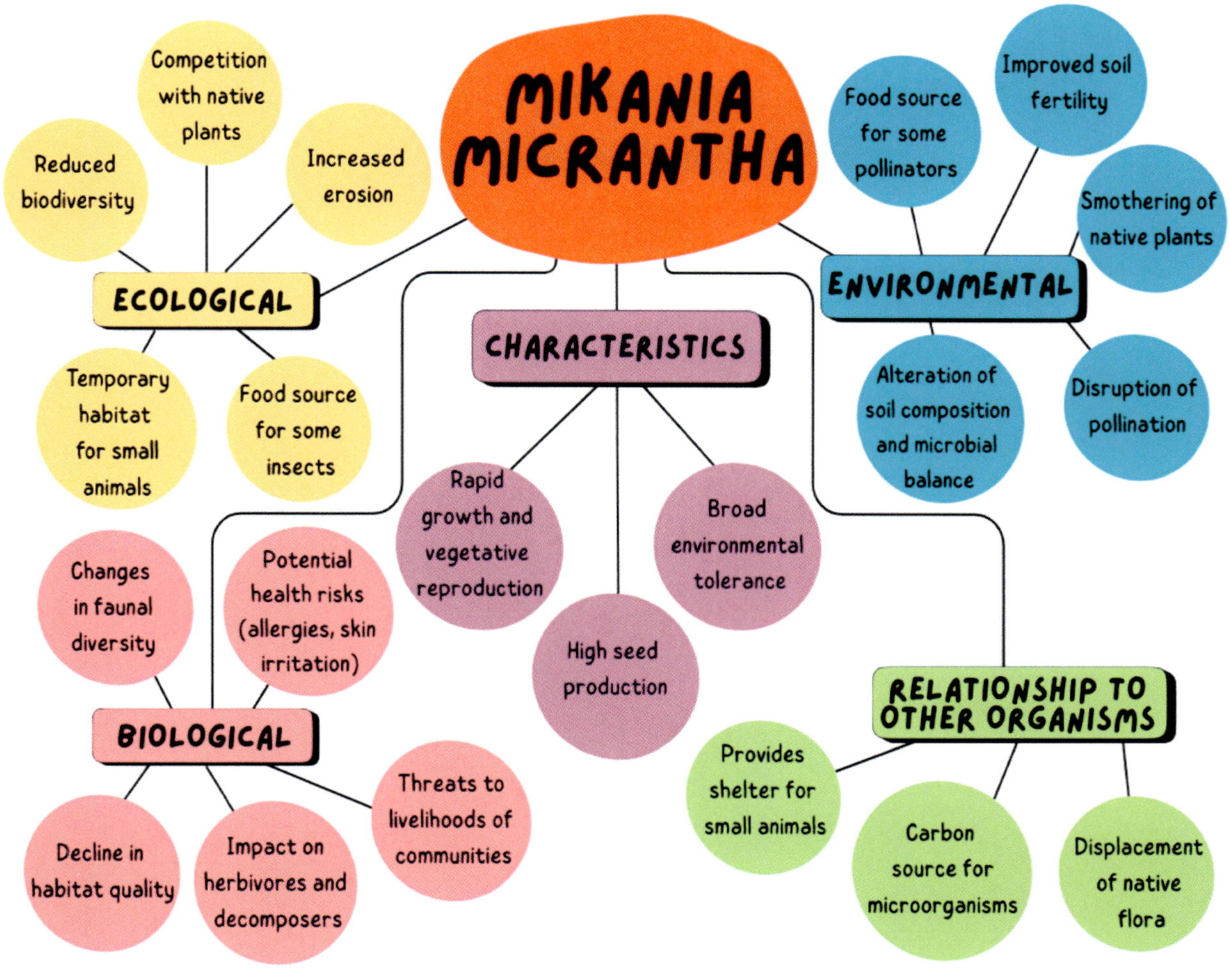

Fig. 8 Concept Map of the invasive species *Mikania micrantha* (Aguila, DRC)

The concept map presents *Mikania micrantha* as a highly invasive vine whose success is driven by rapid growth, vigorous vegetative reproduction, high seed output, and broad environmental tolerance. These characteristics enable it to spread quickly across diverse habitats. Ecologically, the species poses major threats by competing with native plants, reducing biodiversity, increasing erosion, and altering faunal diversity. Although it may offer temporary habitat or limited food sources for some insects, these benefits are overshadowed by its overall disruption of ecosystem balance. Biologically, *Mikania micrantha* contributes to declining habitat quality, affects herbivores and decomposers, and can even cause health issues in humans, such as allergies and skin irritation. It can also impact local communities whose livelihoods depend on healthy, functioning ecosystems. From an environmental standpoint, the vine alters soil composition and microbial communities, disrupts pollination, and often smothers native vegetation, though it may occasionally improve soil fertility or serve as a resource for certain pollinators. In its interactions with other organisms, *Mikania micrantha* can displace native flora, provide shelter for small animals, and act as a carbon source for microorganisms. Overall, the concept map emphasizes how this invasive species influences ecological, biological, and environmental systems in interconnected ways, ultimately illustrating its predominantly detrimental impact on natural ecosystems.

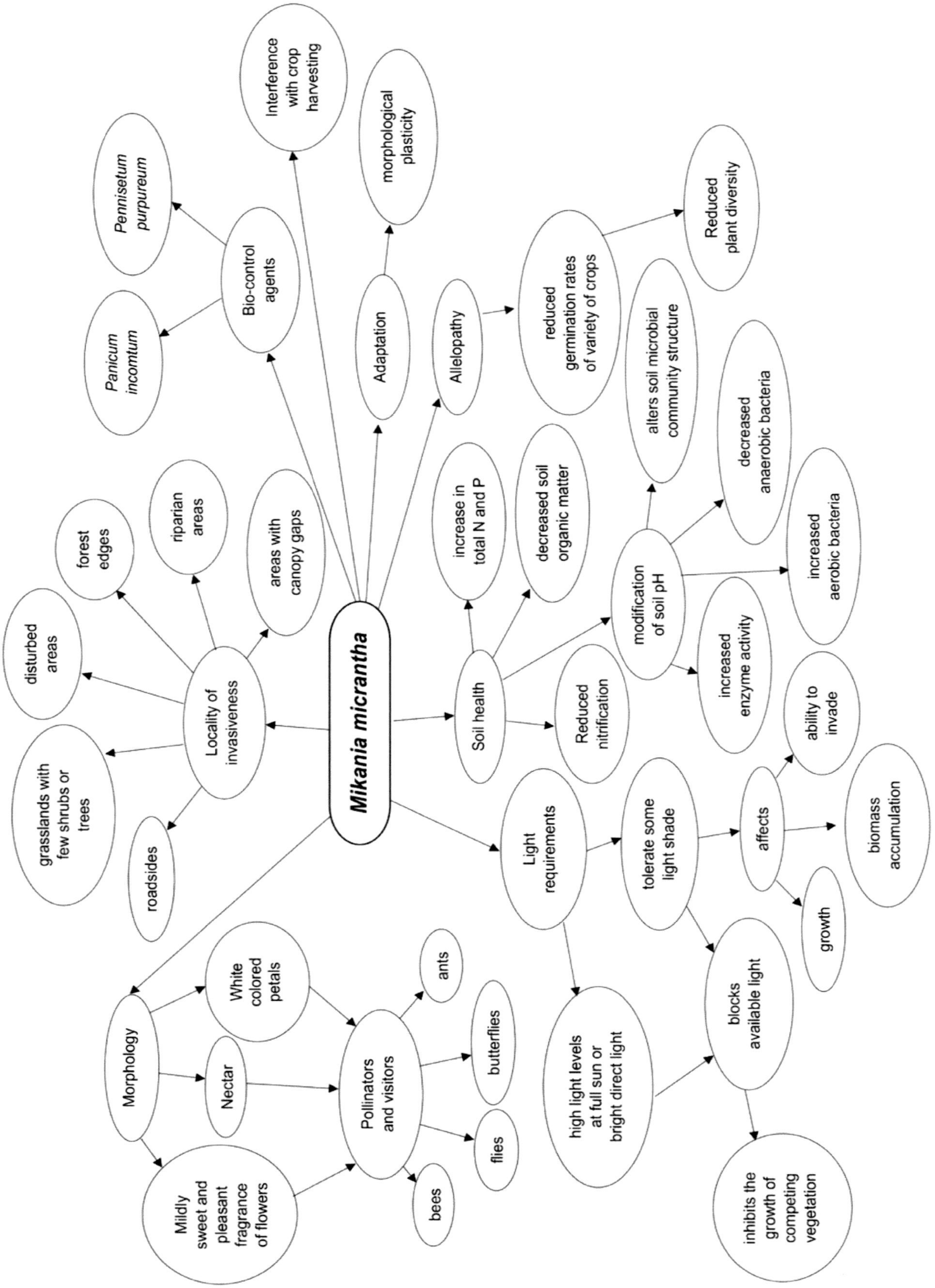

Fig. 9 Invasive Species Concept Map of *Mikania micrantha* (Bautista, RID)

Shown on the previous page is the generated concept map of *M. micrantha*'s invasiveness. Through the species' morphological characteristics, such as its ability to release a mildly sweet, pleasant fragrance from its flowers, aided by its white petals and nectar, pollinators and visitors were able to pollinate it and allow its rapid spread. Some pollinators and visitors of *M. micrantha* are bees, flies, butterflies, and ants. Additionally, environmental conditions have increased this species' potential to invade various localities, as it can be found on roadsides, in grasslands with few shrubs or trees, in disturbed areas, at forest edges, in riparian areas, and in areas with canopy gaps. The light requirements for *M. micrantha* to thrive are primarily high light levels, such as full sun or bright direct light; however, it can tolerate some light shade. *M. micrantha* growing in some shaded areas could affect its growth, ability to invade, and decrease biomass accumulation. Both growing in direct light and shaded light block the available light for their host species, thereby inhibiting the growth of its competing vegetation. Despite these concerns, *M. micrantha* has managed to adapt to various environmental conditions, particularly due to its morphological plasticity.

Regarding its effects on soil health, it has been shown to reduce nitrification, increase total N and P available to plants, decrease soil organic matter, and alter soil pH. Altering the soil pH can alter soil microbial community structure, increase aerobic bacteria, decrease anaerobic bacteria, and increase enzyme activity. With the species' allelopathic potential, reduced germination rates of various crops have been reported. Overall, its impacts have disrupted crop harvesting, reducing yields of economically important crops.

Despite these invasive concerns with *M. micrantha*, control and eradication measures were continuously being tested and studied. Some of these biocontrol agents include *Panicum incomtum* and *Pennisetum purpureum*. On the other hand, other control measures have been proven to help limit its ecological dominance but not completely eradicate *M. micrantha*.

References

Alcalde, G.T., & Buot, I.E. (2023). *Mikania micrantha* Kunth (Asteraceae) as a potential invasive in terrestrial and wetland ecosystems. *J. Wetlands Biodiversity* 13: 69-82

Ali Khan, M., El-Kersh, D. M., Islam, M. S., Ara Khan, S., Kamli, H., Sarkar, C., Bhuia, M. S., Islam, T., Chandra Shill, M., Gobe, G. C., Sönmez Gürer, E., Setzer, W. N., Sharifi-Rad, J., & Torequl Islam, M. (2023). *Mikania micrantha* Kunth: An Ethnopharmacological Treasure Trove of Therapeutic Potential. *Chemistry & Biodiversity*, *20*(11), e202300392. https://doi.org/10.1002/cbdv.202300392

Banerjee, A. K., Mukherjee, A., Guo, W., Liu, Y., & Huang, Y. (2019). Spatio-Temporal Patterns of Climatic Niche Dynamics of an Invasive Plant Mikania micrantha Kunth and Its Potential Distribution Under Projected Climate Change. *Frontiers in Ecology and Evolution*, *7*, 477754. https://doi.org/10.3389/fevo.2019.00291

Bora Rahman Atiqur, Babu Sunil Dasi, Chetry Sita, Kalita Sontara (2023). Management Options of *Mikania micrantha*: A Review. *Agricultural Reviews*. *44*(3): 393-396. doi: 10.18805/ag.R-2201.

Cock, M. J. W. (1982). Potential biological control agents for *Mikania micrantha*hbk from the Neotropical region. *Tropical Pest Management*, *28*(3), 242–254. https://doi.org/10.1080/09670878209370717

Day, M. (2012). *Biological Control of Weeds in Australia* (M. Julien, R. McFadyen, & J. Cullen, Eds.; pp. 368–372). CSIRO. https://books.google.com/books?hl=en&lr=&id=PN8xW4uuf3EC&oi=fnd&pg=PA368&dq=mikania+micrantha&ots=dnmx3Mvwbm&sig=Vky3B8U9tBPxURukjFzVGxKjjXQ

Day, M. D., Clements, D. R., Gile, C., Senaratne, W. K. A. D., Shen, S., Weston, L. A., & Zhang, F. (2016). Biology and Impacts of Pacific Islands Invasive Species. 13. *Mikania micrantha* Kunth (Asteraceae). *Pacific Science*, *70*(3), 257–285. https://doi.org/10.2984/70.3.1

Day, M.D., Kawi, A., Kurika, K., Dewhurst, C.F., & Waisale, S. (2012). *Mikania micrantha* Kunth (Asteraceae) (milea-minute): Its distribution and physical and socioeconomic impacts in Papua New Guinea, *Pac. Sci.* 66: 213-223.

Hong, L., Liu, M.J., Shen, H., Ye, W.H., & Cao, H.L. (2010). Reproductive allocation of biomass in *Mikania micrantha* populations from two habitats, [In Chinese.] *J. Wuhan Bot. Res.* 28: 598–605.

iNaturalist. (n.d.). *Mile-a-Minute (Mikania micrantha)*. INaturalist. https://www.inaturalist.org/taxa/165389-Mikania-micrantha

Legarda, B. J. (1999). *After the Galleons: Foreign Trade, Economic Change and Entrepreneurship in the Nineteenth-century Philippines*. Madison: University of Wisconsin-Madison.

Li, Y., Li, J., Li, Y., Wang, X., & Cao, A. (2013). Antimicrobial Constituents of the Leaves of *Mikania micrantha* H. B. K. *PLoS ONE*, 8(10), e76725. https://doi.org/10.1371/journal.pone.0076725

Lowe, S., Browne, M., Boudjelas, S., & De Poorter, M. (2001). *100 of the world's worst invasive alien species: A selection from the global invasive species database*, IUCN/SSC.

National Parks Board. (2025, September 2). *NParks | Mikania micrantha*. Www.nparks.gov.sg. https://www.nparks.gov.sg/florafaunaweb/flora/1/4/1451

Pelser, P.B., J.F. Barcelona & D.L. Nickrent (eds.). 2011 onwards. Co's Digital Flora of the Philippines. www.philippineplants.org

Poudel, M., Adhikari, P., & Thapa, K. (2019). Biology And Control Methods Of The Alien Invasive Weed *Mikania Micrantha*: A Review. *Environmental Contaminants Reviews*, *2*(2), 06-12. https://doi.org/10.26480/ecr.01.2019.06.12

Puzari K.C., Bhuyan R.P., Dutta P., Nath H.K.D. (2010). Distribution of *Mikania* and its economic impact on tea ecosystem of Assam, *Indian Journal of Forestry 33*(1): 71- 76.

Royal Botanic Gardens, Kew. (2021). *Mikania micrantha Kunth | Plants of the World Online | Kew Science*. Plants of the World Online. https://powo.science.kew.org/taxon/urn:lsid:ipni.org:names:161051-2

Sheam, M., Haque, Z., & Nain, Z. (2020). Towards the antimicrobial, therapeutic and invasive properties of *Mikania micrantha* Knuth: a brief overview. *Journal of Advanced Biotechnology and Experimental Therapeutics*, *3*(2), 92. https://doi.org/10.5455/jabet.2020.d112

Shen, S., Xu, G., Zhang, F., Jin, G., Liu, S., Liu, M., Chen, A., and Zhang, Y. (2013). Harmful Effects and Chemical Control Study of *M. micrantha Micrantha* H.B.K. in Yunnan, Southwest China. African Journal of Agricultural research. 8(44): 5554-61.

Tripathi, R. S., Khan, M. L., & Yadav, A. S. (2012). Biology of *Mikania micrantha* H.B.K.: a review. *Invasive Alien Plants: An Ecological Appraisal for the Indian Subcontinent*, 99–107. https://doi.org/10.1079/9781845939076.0099

Tropical Plants Database, Ken Fern. (2025). *Mikania micrantha - Useful Tropical Plants*. Theferns.info. https://tropical.theferns.info/viewtropical.php?id=Mikania+micrantha

Watson, G.A., Wong, P.W., & Narayanan, R. (1964). Effects of cover plants on soil nutrient status and on growth of *Hevea*. IV. Leguminous creepers compared with grasses, *Mikania cordata* and mixed indigenous covers on four soil types, *Journal of the Rubber Research Institute of Malaya* 18: 123-145.

Widjaja, E.A., & Tjitrosoedirdjo, S.S. (1991). The development of weeds under *Dendrocalamus asper* (Schultz. F.) Back. ex Heyne plantation in Lampung, Sumatra, Indonesia, Proceedings, 13th Asian-Pacific Weed Science Society Conference Taipei, Taiwan, *Asian-Pacific Weed Science Society* 1: 225-227.

Wong, P.W. (1964). Evidence for the presence of growth inhibitory substances in *Mikania cordata* (Burm. f.) B.L. Robinson, *J. Rubber Res. Inst. Malaya* 18: 231-242.

Yang, Q.H., Ye, W.H., Deng, X., Cao, H.L., Zhang, Y., & Xu, K.Y. (2005). Seed germination eco-physiology of *Mikania micrantha* H.B.K., *Bot. Bull. Acad. Sin.* (Taipei) 46: 293–299.

Zhang, L. Y., Ye, W. H., Cao, H. L., & Feng, H. L. (2004). *Mikania micrantha* H. B. K. in China - an overview. *Weed Research*, *44*(1), 42–49. https://doi.org/10.1111/j.1365-3180.2003.00371.x

D

Universiti Tun Hussein Onn Malaysia

Creeping Wood Sorrel (*Oxalis corniculata*): An Invasive Groundcover

Yung Li Zhi & Teu Yoke Yan

A. Invasive Species Name

Oxalis corniculata, commonly known as creeping wood sorrel or yellow wood sorrel, is a widespread perennial weed known for its extremely invasive nature. Its current global distribution reflects a long history of accidental spread through human activity and its remarkable ecological adaptability. Although wood sorrel is often associated with lawns, gardens, and roadsides, it is capable of thriving in a variety of disturbed and semi-natural environments, raising concerns about its ecological and economic impacts

B. Native Range and Method of Introduction

Due to its long history of association with human-inhabited areas, the exact origin of wood sorrel remains uncertain. However, it is generally believed to be native to Europe or parts of South and Southeast Asia, eventually becoming naturalized in tropical and temperate regions (NParks, 2025). The species likely spread globally through various pathways, such as contaminated soil, nursery plants, lawns, and crops. Its tiny seeds are easily dispersed by wind, water, animals, and human footwear, while its explosive seed dispersal mechanism allows it to efficiently colonize local areas (NParks, 2025). Collectively, these characteristics have facilitated the species' successful introduction and widespread colonization of urban, agricultural, and disturbed environments.

Morphologically, *O. corniculata* is a low-growing herbaceous plant characterized by stolons, roots that form at the stem nodes, compound leaves with three leaflets, and small yellow flowers. It is highly tolerant of a wide range of environmental conditions, capable of growing in full sun, partial shade, and even compacted or poor soil (NParks, 2025). Rashid et al. (2020) applied Grime's Competitive-Stress-tolerant-Ruderal (CSR) triangle model to classify wood sorrel as a Competitive Ruderal (C-R) garden weed, meaning it performs well in moderately productive but frequently disturbed habitats. Studies have shown that *O. corniculata* has the longest reproductive period among its group, reproducing both sexually (by seed) and asexually throughout the year (Rashid et al., 2020). This distinguishes it from other closely related species that reproduce only asexually

Research highlights the presence of a variety of phytochemicals, antioxidants, and other metabolites in wood sorrel, reflecting its strong metabolic diversity (Groom et al., 2019). A recent phytochemical investigation found an abundance of antioxidant compounds and other bioactive metabolites in *O. corniculata*, demonstrating its potential as both a medicinal plant and an ecologically robust species (Zhong et al., 2025). Furthermore, a systematic review by Mekap et al. (2022) emphasizes the plant's pharmacological potential, noting the presence of flavonoids, glycosides, and other compounds with anti-inflammatory, anxiolytic, and therapeutic properties. Taken together, these studies provide further evidence that wood sorrel is a potent invasive species, characterized by both ecological tolerance and biochemical complexity. Its metabolic adaptability and efficient reproductive strategy help explain its ability to invade effectively and survive in a variety of stressful environments.

C. Ecological and/or Economic Impacts

The ecological impact of wood sorrel is primarily linked to its ability to form dense mats. The plant creeps along the ground and roots at the nodes, allowing it to quickly cover open soil. This growth habit inhibits the germination and growth of native groundcover species (NParks, 2025). Furthermore, its tolerance of frequent mowing, foot traffic, drought, and soil compaction enables the species to thrive in environments such as urban parks and lawns. Rashid et al. (2020) observed that wood sorrel remains active year-round, with growth slowing only during very cold winters. Consequently, it can rapidly re-establish itself in disturbed soil before other species have a chance to recover. The combination of rapid vegetative spread and explosive seed dispersal allows *O. corniculata* to outcompete slow-growing native plants, reducing plant diversity in disturbed habitats.

In addition, the tolerance of wood sorrel (Oxalis corniculata) to polluted soil—and its ability to accumulate contaminants—is of particular interest. In pot experiments where wood sorrel was grown in mercury-contaminated soil, the addition of chelating agents such as sodium thiosulfate or ethylenediaminetetraacetic acid (EDTA) significantly enhanced mercury uptake in the plant's above-ground parts. This treatment also altered the rhizosphere microbial community, indicating that the plant's interactions with the soil are complex and dynamic (Liu et al., 2018). These results align with the broader findings of a recent phytoremediation review, which suggests that metal-chelating compounds produced by plants themselves are key to the accumulation or stabilization of heavy metals (Parida et al., 2024).

Economically, wood sorrel has become a problematic weed in landscaping, nursery production, and agriculture. It often contaminates potting soil and nursery containers, making it difficult to maintain weed-free conditions for ornamental plants (NParks, 2025). In agricultural fields, it competes with vegetable seedlings and other low-growing crops, which can reduce yields and necessitate repeated weeding or herbicide use. Controlling the plant with herbicides is challenging because its creeping growth habit reduces spray coverage, and surviving stem nodes can easily regenerate new shoots. As a result, long-term management requires frequent attention and increases costs for farmers, landscapers, and land managers. Overall, the plant's adaptability, extended reproductive season, and tolerance to environmental stress demonstrate why *O. corniculata* is viewed as both ecologically disruptive and economically burdensome.

D. Control and Eradication Measures

Because *Oxalis corniculata* possesses strong regenerative abilities, a year-round reproductive cycle, and stems that root at the nodes, comprehensive management measures are necessary to control its growth (Rashid et al., 2020). Mechanical control is often used as a front-line management tool, particularly in gardens and landscapes. If executed properly, manual removal can be effective, provided that all roots and stolons are extracted, as *Oxalis corniculata* can easily regenerate from small remaining fragments (NParks, 2025). Mechanical methods such as plastic film mulching and soil solarization also help inhibit growth by limiting light exposure and reducing seed viability. In lawn areas, regular mowing can reduce seed production, but it cannot completely eradicate established plants. Since *Oxalis corniculata* grows rapidly in disturbed soils, mechanical control alone is often insufficient unless soil health and ground cover management are improved simultaneously

Currently, there are no approved classical biological control agents for *Oxalis corniculata*, so biological control options are limited. This is mainly because other Oxalis species have significant economic or horticultural value, making it difficult to develop host-specific biological control methods without risking collateral damage. However, indirect biological control methods can support management efforts. Maintaining dense and healthy lawns or ground cover can reduce the likelihood of Oxalis colonization, as strong competition for nutrients inhibits its early growth (Rashid et al.,

2020). *Oxalis corniculata* is known to be associated with various microorganisms and endophytes, including heavy metal-tolerant endophytes, which may partly explain its adaptability in disturbed or polluted sites (Groom et al., 2019; Rashid et al., 2020). While not a standalone strategy, improving soil structure and microbial diversity through organic amendments can enhance the competitive advantage of local species over invasive weeds like Oxalis sp

Chemical control is typically used when mechanical removal is impractical or when the infestation is widespread. Non-selective herbicides, such as glyphosate, are widely used for spot treatments along roads, fences, and non-cultivated areas, though care must be taken to avoid damaging surrounding vegetation. In lawn systems, selective broad-leaf herbicides containing 2,4-D, dicamba, and alachlor are usually used to control wood sorrel without damaging turfgrass (NParks, 2025). These herbicides are most effective when applied to young, vigorous plants. For long-term control, pre-emergent herbicides such as oxadiazon or pendimethalin can be used to prevent seed germination, especially in nurseries and landscape flower beds where frequent soil disturbance exposes seed banks. Chemical control must be combined with cultural practices—such as reducing soil disturbance, improving drainage, and maintaining dense vegetation coverage—because wood sorrel can easily colonize disturbed or bare soil through seed dispersal and vegetative fragmentation (Rashid et al., 2020). generally speaking, it is difficult to effectively eradicate this invasive species with a single method; the most effective approach is comprehensive management combining mechanical, ecological, and chemical strategies

E. Original Field Photos

Fig. 1 *Oxalis corniculata*, including flower, leaf shape, and stolon

F. Invasive Species Concept Map

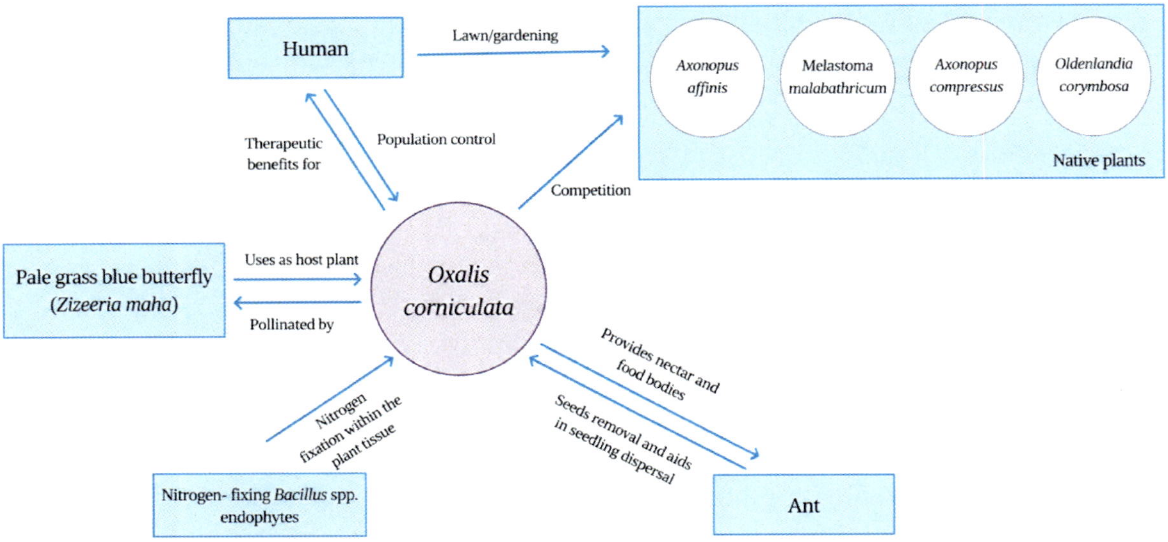

There is a significant relationship between *Oxalis corniculata* and endophytic nitrogen-fixing bacteria, particularly Bacillus sp., which is often observed in nutrient-poor soil. These bacteria form a symbiotic relationship with *Oxalis corniculata*, providing the plant with essential nitrogen (Jooste et al., 2019). This symbiosis gives the species a competitive advantage over native ground vegetation that relies solely on soil nitrogen. Furthermore, these nitrogen-fixing endophytes improve the plant's stress resistance and promote rapid regeneration after removal, contributing to its continued survival and invasion in lawns, gardens, and agricultural landscapes.

In addition, *Oxalis corniculata* serves as the larval host plant for the pale grass blue butterfly (Zizeeria maha). The larvae of this butterfly feed exclusively on the leaves of *Oxalis corniculata*. Consequently, changes in the plant's physiology or abundance—driven by competition or invasive dominance—can directly influence the butterfly's survival and distribution (Ko Sakauchi et al., 2021). This host-plant relationship represents a key biotic interaction between an invasive plant and a native insect species, which may indirectly facilitate the invasion process.

Furthermore, ants play a crucial role in seed dispersal through a process known as myrmecochory. The seeds of *Oxalis corniculata* possess nutrient-rich appendages called elaiosomes that attract ants. When ants encounter these seeds, they carry them back to their nests to consume the elaiosome (Karnish, 2024). Once the elaiosome is consumed, the ants discard the intact seed either inside or outside the nest. This places the seed in a safe, nutrient-enhanced environment away from the parent plant, allowing for successful germination. This ant-mediated dispersal reduces competition among seedlings and promotes the spread of *Oxalis corniculata* in disturbed habitats.

Finally, *Oxalis corniculata* is widely utilized by humans for its nutritional and therapeutic benefits. According to Sarkar et al. (2020), the plant contains essential phytochemicals like niacin, vitamin C, and β-carotene, and is used to treat a variety of ailments, such as fever, skin infections, and stomach issues. In some rural areas, it is even used as a seasoning or green leafy vegetable. However, despite these benefits, *Oxalis corniculata* exhibits allelopathic properties that can damage horticultural crops and native groundcover. Ironically, humans often inadvertently promote its growth by maintaining the very lawns and gardens where it thrives. Case studies show that while *O. corniculata* coexists with lawn grass, its rapid growth allows it to crowd out other ground cover plants; it was found in up to 80% of investigated garden plots (Masita et al., 2023). Therefore, human management of these habitats has inadvertently facilitated the plant's invasion, allowing it to inhibit and compete with desirable horticultural species.

References

Groom, Q. J., Van der Straeten, J., & Hoste, I. (2019). The origin of Oxalis corniculate L. PeerJ, 7, e6384. https://doi.org/10.7717/peerj.6384

Jooste, M., Roets, F., Midgley, G. F., Oberlander, K. C., & Dreyer, L. L. (2019). *Nitrogen-fixing bacteria and Oxalis – evidence for a vertically inherited bacterial symbiosis. BMC Plant Biology, 19*(1). https://doi.org/10.1186/s12870-019-2049-7

Karnish, A. (2024). *Seed Dispersal by Ants: A Primer. International Journal of Plant Sciences, 185*(5), 403–411. https://doi.org/10.1086/730787

Ko Sakauchi, Taira, W., & Otaki, J. M. (2021). *Metabolomic Response of the Creeping Wood Sorrel Oxalis corniculata to Low-Dose Radiation Exposure from Fukushima's Contaminated Soil. Life, 11*(9), 990–990. https://doi.org/10.3390/life11090990

Liu, Z., Wang, L., Ding, S., & Xiao, H. (2018). Enhancer assisted-phytoremediation of mercury-contaminated soils by Oxalis corniculata L., and rhizosphere microorganism distribution of Oxalis corniculata L. Ecotoxicology and Environmental Safety, 160, 171–177. https://doi.org/10.1016/j.ecoenv.2018.05.041

Masita, R., Kartikasari, N., Lukiati, B., & Sulisetijono. (2023). Exploration of Oxalis sp. growth inhibition potential based on the diversity and distribution of ground vegetation. AIP Conference Proceedings, 2634, 020118. https://doi.org/10.1063/5.0113882

Mekap, S. K., Mishra, S. K., Panda, P. K., Panda, S. S., & Sarangi, D. K. (2022). *A Systematic Review On Oxalis Corniculata Linn. A Crop Field Weed With Promising Pharmacological Activities. Asian Journal of Pharmaceutical and Clinical Research, 15*(8), 4–8. https://doi.org/10.22159/ajpcr.2022.v15i8.45017

NParks (18 November 2025) Oxaliscorniculata National Parks Flora & Fauna Web, www.nparks.gov.sg. https://www.nparks.gov.sg/florafaunaweb/flora/2/2/2292

Parida, T., Riyazuddin, S., Kolli, S. K., Chakraborty, A., Srinivas, N., Kundu, P., Bhattacharya, S., Seth, C. S., & Biswas, J. K. (2024). *Phytoremediation of heavy metal(loid)s with integral involvement of the endogenous metal-chelators: present state-of-the-art and future prospect. Discover Environment, 2*(1), 139. https://doi.org/10.1007/s44274-024-00170-x

Rashid, A., Cheema, T. A., Khan, A. U. H., Zahid, F., Bashir, Z., Mirza, S. A., & Ali, S. (2020). *Strategic exploration of potential growth for the selection of niches for Oxalis spp. JAPS: Journal of Animal & Plant Sciences, 30*(6), 1633

Sarkar, T., Ghosh, P., Poddar, S., Choudhury, S., Sarkar, A., & Chatterjee, S. (2020). *Oxalis corniculata Linn. (Oxalidaceae): A brief review. Journal of Pharmacognosy and Phytochemistry, 9*(4), 651–655. https://doi.org/10.22271/phyto.2020.v9.i4i.11777

Zhong, T., He, J., Zhao, H., Tan, C., Zhou, W., Wu, C., & Kang, J. (2025). *Oxalis corniculata L. As a Source of Natural Antioxidants: Phytochemistry, Bioactivities, and Application Potential. Antioxidants, 14*(11), 1352. https://doi.org/10.3390/antiox14111352

Soap Bush (*Miconia crenata*): Shrub Invasion in Disturbed Sites

Janice Lee Jing Yan & Tan Yan Zi

A. Invasive Species Name

Miconia crenata (a synonym for *Clidemia hirta*), a member of the Melastomataceae family, is classified as a highly invasive tropical shrub. This perennial woody plant typically reaches up to 3 metres in height, with stems covered in distinctive red bristles. Its leaves are simple, opposite, and attached by petioles. The leaves are hairy and ovate-to-oblong with crenate (scalloped) margins, and their most distinguishing feature is the five major veins that radiate from the base. Additionally, the plant produces white flowers with five petals (Global Invasive Species Database, 2025).

The flowers grow from axillary buds and are arranged in small clusters at the branch tips. The fruits are round, pulpy, dark-blue to purple berries, with each one containing between 200 and 900 seeds (Wester & Wood, 1977; Dewalt, Denslow, & Hamrick, 2004). The shrub reproduces primarily through sexual reproduction, with seeds dispersed by fruit-eating birds and other animals, but it can also reproduce asexually through apomixis (Global Invasive Species Database, 2025).

B. Native Range and Method of Introduction

Miconia crenata is native to the tropical Americas, including South America, Central America, and the Caribbean (DeWalt, Denslow, & Johnson, 2022). In its native range, it typically appears at low densities in early-successional plant communities, establishing it as a pioneer species in secondary rainforests. However, it has since spread throughout the Old World Tropics, becoming a serious pest in regions such as Oceania, Southeast Asia, and the Indian subcontinent (Global Invasive Species Database, 2025). The plant was first collected on the island of Oahu in 1949 and was identified as a hazardous species by 1957. Later, a 1971 seed bank study led by P.S. Ashton first detected the presence of Clidemia hirta seeds at the Pasoh Forest Reserve in Peninsular Malaysia (Peters, 2001). From these points of introduction, it spread quickly. For instance, in Hawaii, the infestation expanded throughout the island chain, with repeated, smaller infestations documented on all major islands except Kahoolawe and Niihau during the 1970s and 1980s.

Humans played a major role in the initial long-distance spread of *M. crenata* from its native range, either accidentally as a contaminant or intentionally. In some cases, *Miconia crenata* was purposefully intro-duced as an ornamental plant. For example, it was cultivated at the Wahiawa Botanic Garden in Hawaii, where it was first recorded in 1941 and initially considered a promising, non-spreading species. In another instance, it is believed that coffee planter William Parr introduced seedlings to Fiji sometime before 1890 (Pyšek, P. et al., 2011).

The rapid expansion of *M. crenata* once introduced is driven by its reproductive success and association with disturbed habitats. A mature plant can produce more than 500 blue-black berries annually, each containing over 100 tiny seeds. The seeds are dispersed locally by frugivorous birds that consume the fruit, as well as by other animals and human activity. Feral pigs are also known to spread the seeds through their feces after eating the fruit.

As an early colonizer, *Miconia crenata* thrives in disturbed areas such as plantations, roadsides, pastures, and forest gaps. In many regions, including Hawaii and Peninsular Malaysia, ground disturbance created by feral pigs facilitates its successful establishment and invasion into natural forests. Furthermore, the shrub benefits from a phenomenon known as "enemy release" in its introduced range. In places like Hawaii, it is freed from the heavy pressures of insect herbivores and fungal pathogens found in its native habitat, which allows it to grow more vigorously and exhibit greater shade tolerance (Richardson, 2011).

C. Ecological and/or Economic Impacts

Clidemia hirta is a fast-growing pioneer shrub that thrives in disturbed, high-light environments where it competes with native seedlings for light, space, and nutrients. Its primary ecological impact stems from its ability to form dense thickets that block the regeneration of native flora. In Malaysia, for example, its dominance in tree-fall gaps has been shown to outcompete native canopy specialists, disrupting natural forest succession and slowing the transition to mature rainforest (Ho, C.-C et al., 2023).

In addition to direct competition, *Clidemia hirta* also vies with native plants for the attention of seed-dispersing birds, which can reduce the reproductive success of native fruiting species. Over time, the colonization and overshading by dense thickets of *Clidemia hirta* not only displaces native plants but can also lead to soil degradation and reduced fertility.

D. Control and Eradication Measures

For small infestations of *Miconia crenata*, hand-pulling seedlings is an effective removal method, particularly for plants less than 3 metres tall (Wittenberg & Cock, 2001). For mature plants, which have extensive root systems and a high reproductive rate, digging them up is the preferred method. This approach eliminates the plant's ability to reproduce and avoids the need for repeated cutting or chemical treatments. For both methods, it is critical to remove the entire root system to prevent regrowth. However, mechanical control does not affect the soil seed bank, so ongoing monitoring is essential. Newly emerging seedlings must be removed for 5 to 10 years after the initial treatment to prevent reinvasion (Wittenberg & Cock, 2001).

Several biological control agents have been tested for *Miconia crenata*. The first agent released in Hawaii in 1953 was *Liothrips urichi*, a species of thrips (Gassmann et al., 2009). This insect feeds on plant sap, which can cause terminal leaves to drop and occasionally lead to plant mortality. However, its effectiveness is limited, as it does not significantly impact flowering or berry production and is only active in sunny or partly sunny areas (Conant, 2002; Reimer & Beardsley, 1989).

Other agents have been largely unsuccessful. The pyralid leaf-roller, *Ategumia matutinalis*, was heavily parasitized by four species of hymenopterans, resulting in high larval mortality and poor establishment in the field (Conant, 2002). The beetle *Lius poseidon* acts as a dual-stage defoliator, with larvae mining the leaves and adults feeding on them, potentially complementing the damage caused by thrips (Conant, 2002). Finally, larvae of the noctuid moth *Antiblemma acclinalis* roll and feed within the leaves, but this species is rarely seen, suggesting that other biotic factors may be interfering with its establishment.

Local studies in Southeast Asian rubber plantations, using a droplet application technique, have shown that *Miconia crenata* has a low susceptibility to common herbicides like triclopyr, dichlorprop, glyphosate, and 2,4-D, suggesting a relative resistance. Despite this, Triclopyr is still used for control, often in combination with cut-stump treatments to inhibit regrowth after cutting. In Hawaii, spraying a cut stump with a 50% triclopyr amine solution has been shown to prevent coppicing. For smaller plants, a 1-2% triclopyr amine foliar spray has also proven effective in areas like the Kipahulu Valley, Maui (Motooka et al., 2003).

Fluroxypyr is another herbicide used to manage *Miconia crenata* (Department of Agriculture, Malaysia [DOA], 2025). As a synthetic auxin, it mimics natural plant growth hormones and is effective for the post-emergence control of broadleaf weeds and woody plants. After being absorbed by the leaves, the herbicide travels to the plant's meristematic tissues, where it causes unregulated cell growth—visible as twisted leaves and elongated stems—which ultimately leads to plant death. Common application methods include high-dose foliar spraying and cut-stump or basal bark treatments.

E. Original Field Photos

Fig. 1 Field photograph of *Miconia crenata* in the disturbed area around student residential area in Universiti Tun Hussein Onn Malaysia, Johor, Malaysia (top) and Gunung Belumut Forest Reserve, Johor, Malaysia (bottom)

F. Invasive Species Concept Map

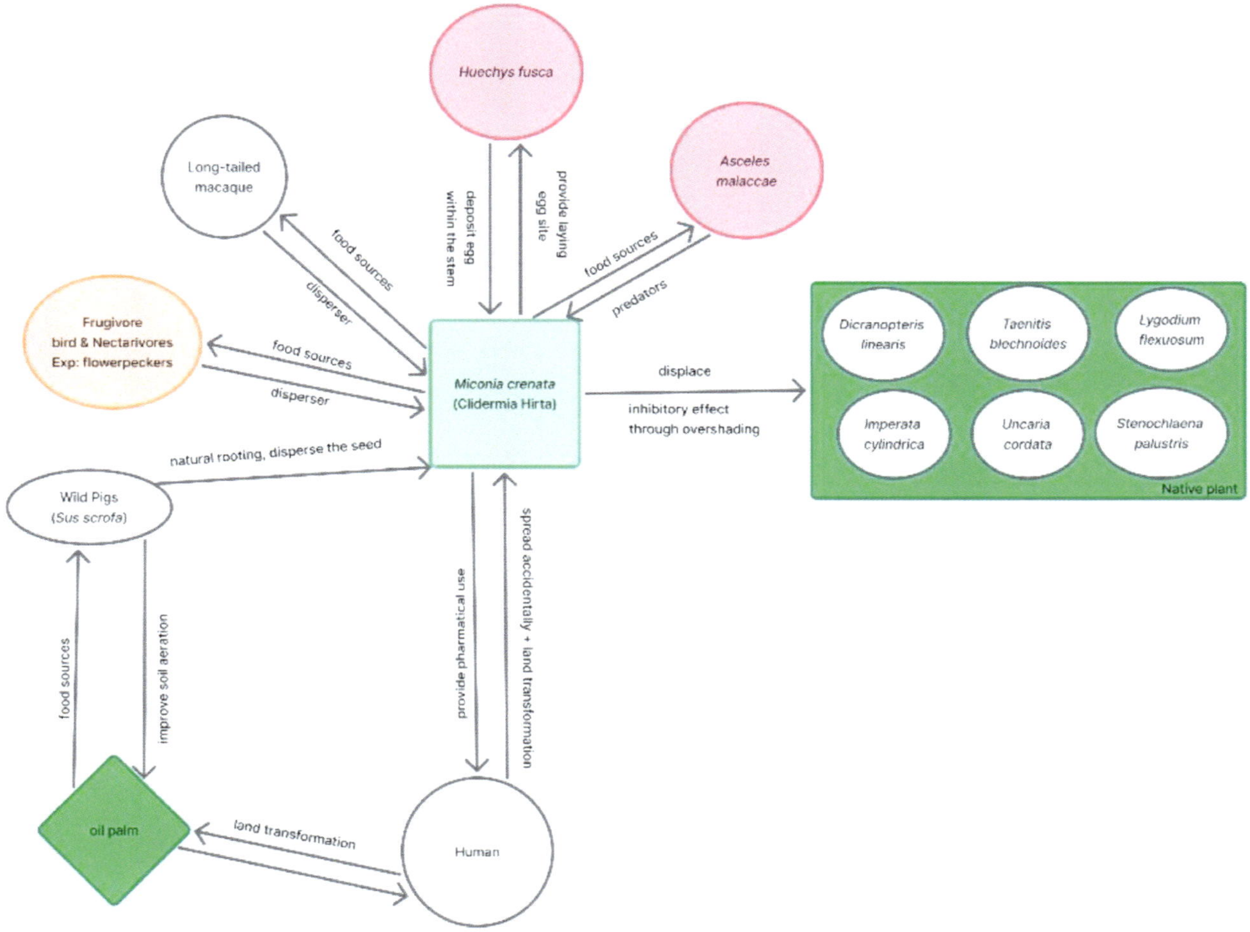

The concept map illustrates how *Miconia crenata* interacts with animals, plants, and humans, forming a complex ecological web. As omnivores, wild pigs consume a varied diet and disturb the soil surface while foraging, using their snouts to overturn and loosen it. This activity often leaves the soil bare and uneven, providing ideal colonization sites for the invasive *Miconia crenata*. Additionally, wild pigs facilitate the plant's spread by dispersing seeds caught on their fur or passed through their droppings. Humans also contribute to the spread of *M. crenata* through landscape transformation for agriculture, such as oil palm cultivation, and through road development. These activities fragment habitats, creating disturbed edges where the plant can easily establish itself. As a result of these disturbances, the rapid growth of *Miconia crenata* displaces native understory and forest floor plants in Dipterocarp forests, including species like *Dicranopteris linearis*, *Taenitis blechnoides*, and *Imperata cylindrica* (Wahyuni et al., 2016).

A key interaction occurs between flowerpeckers and *Miconia crenata*, where the plant serves as a food source and the bird acts as a primary seed disperser. The plant produces small, purplish berries containing hundreds of tiny seeds. When flowerpeckers consume the fruit, the seeds pass through their digestive system unharmed and are dispersed in their droppings over long distances. This process helps the invasive plant spread to new areas it could not otherwise reach (Hss, 2023).

Similarly, long-tailed macaques are also important seed dispersers. As opportunistic omnivores, they feed on wild fruits in the same forest edges, trails, and secondary growth areas where *Miconia crenata* grows abundantly. Furthermore, macaques often disturb the soil and vegetation in these habitats, creating ideal microhabitats for the seeds to germinate. In this way, macaques not only spread the seeds but also create suitable conditions for them to grow (Lucas et al., 1991).

Several native insects now use *Miconia crenata* as a food source. Huechys fusca, a cicada species also known as the Black and Golden Cicada, deposits its eggs within the stems of the plant. This demonstrates that *M. crenata* has become integrated into the local ecosystem, serving as a host for native insects (Leong et al., 2011). Another example is Asceles malaccae, a native Malaysian stick insect. As a herbivore, this insect consumes the leaves of various plants, including *Miconia crenata*, and may therefore contribute to minor biological control of the population.

References

Conant, P. (2002). Classical biological control of Clidemia hirta (Melastomataceae) in Hawai'i using multiple strategies. Pacific Cooperative Studies Unit, University of Hawai'i at Mānoa.

Department of Agriculture, Malaysia. (2025). List of Registered Pesticides.

Dewalt, S. J., Denslow, J. S., & Hamrick, J. L. (2004). Biomass allocation, growth, and photosynthesis of genotypes from native and introduced ranges of the tropical shrub Clidemia hirta. Oecologia, 138(4), 521–531. https://doi.org/10.1007/s00442-003-1462-6

DeWalt, S. J., Denslow, J. S., & Johnson, M. T. (2022). Ecology and control of naturalized Melastomataceae. In Springer EBooks (pp. 761–789). https://doi.org/10.1007/978-3-030-99742-7_34

Gassmann, A., Toševski, I., Jović, J., Guazzone, N., & Nguyen, D. (2009). Biological control of buckthorns, Rhamnus cathartica and Frangula alnus: Report 2008–09. Institute for Plant Protection and Environment, Belgrade, Serbia. Retrieved from https://www.lccmr.mn.gov/projects/2007/finals/2007_04i_rpt_buckthorn.pdf

Global Invasive Species Database. (2025). Species profile: Clidemia hirta. Downloaded from http://www.iucngisd.org/gisd/species.php?sc=53 on 01-11-2025.

Ho, C.-C., Bastien Dehaudt, Y.-H, B. P., Tan, H., & Luskin, M. S. (2023). Recolonizing native wildlife facilitates exotic plant invasion into Singapore's rain forests. Biotropica, 55(5), 1033–1044. https://doi.org/10.1111/btp.13251

Hss, A. (2023, July 24). Flowerpeckers that feed on the fruit of Clidemia hirta syn. Miconia crenata. Bird Ecology Study Group. Retrieved from https://besgroup.org/2023/07/24/flowerpeckers-that-feed-on-the-fruit-of-clidemia-hirta-syn-miconia-crenata/

Leong, T. M., Shunari, M., Leong, Y. K., & Foo, S. K. (2011). Records of the Black and Golden Cicada, Huechys fusca Distant, 1892 in Singapore, with natural history observations (Homoptera: Cicadidae: Cicadettinae). Nature in Singapore, 4, 203-211.

Lucas, W., & Corlett, T. (1991). Relationship between the diet of Macaca fascicularis and forest phenology. Folia Primatologica, 57(4), 201-215.

Motooka, P., Castro, L., Nelson, D., Nagai, G., & Ching, L. (2003). Weeds of Hawaii's pastures and natural areas: An identification and management guide. College of Tropical Agriculture and Human Resources, University of Hawai'i at Mānoa.

Peters, H. A. (2001). Clidemia hirta invasion at the Pasoh Forest Reserve: An unexpected plant invasion in an undisturbed tropical forest. Biotropica, 33(1), 60–68. https://doi.org/10.1111/j.1744-7429.2001.tb00157.x

Pyšek, P., Jarošík, V., Chytrý, M., Danihelka, J., Kühn, I., Pergl, J., Tichý, L., Biesmeijer, J. C., Ellis, W. N., Kunin, W. E., & Settele, J. (2011). Successful invaders co-opt pollinators of native flora and accumulate insect pollinators with increasing residence time. Ecological Monographs, 81, 277-293.

Richardson, D. M., & Rejmánek, M. (2011). Trees and shrubs as invasive alien species – a global review. Diversity and Distributions, 17, 788-809.

Reimer, N. J., & Beardsley, J. W. (1989). Effectiveness of Liothrips urichi (Thysanoptera: Phlaeothripidae) introduced for biological control of Clidemia hirta in Hawaii. Environmental Entomology, 18(6), 1141–1146. https://doi.org/10.1093/ee/18.6.1141

Wahyuni, I., Sulistijorini, Setiabudi, Meijide, A., Nomura, M., Kreft, H., Rembold, K., Tjitrosoedirdjo, S. S., & Tjitrosoedirdjo, S. (2016). Distribution of invasive plant species in different land-use systems in Sumatera, Indonesia. Biotropia, 23(2). https://doi.org/10.11598/btb.2016.23.2.534

Wester, L. L., & Wood, H. B. (1977). Koster's curse (Clidemia hirta), a weed pest in Hawaiian forests. Environmental Conservation, 4(1), 35–41.

Wittenberg, R., & Cock, M. J. W. (Eds.). (2001). Invasive alien species: A toolkit of best prevention and management practices. CAB International on behalf of the Global Invasive Species Programme.

Billygoat Weed (*Ageratum conyzoides*): Fast-Spreading Weed of Farms and Roadsides

Muhammad Haziq Asyraaf Abu Bakar & Zahirulhaq Assiddiq Azmi

A. Invasive Species Name

Ageratum conyzoides (Asteraceae), locally known as Rumput Tahi Ayam, is commonly referred to in English as Billygoat weed, Goatweed, Chickweed, or Tropical whiteweed. It is a widely distributed annual herbaceous plant found throughout tropical and subtropical regions. The species is readily distinguished by its characteristic woolly pale-violet to white flowers and a distinctive odour attributed to volatile secondary metabolites (Kumar & Jhariya, 2013). The genus name *Ageratum* is derived from the Greek a, meaning "non," and geras, meaning "age," in reference to its long-lasting flowers.

Ageratum conyzoides is cited as one of the world's most serious invasive weed species because it produces large quantities of seeds, releases poisonous compounds that suppress other plants, and spreads rapidly into areas disturbed by human activity (Sharma et al., 2017). In numerous parts of Asia, including Malaysia, *Ageratum conyzoides* has been identified as a fast-growing weed that poses a threat to agricultural crops, tree plantations, forest edges, and certain roadside areas.

B. Native Range and Method of Introduction

During the European colonisation of America and Africa, *Ageratum conyzoides*, native to Central America, Mexico, and parts of the Caribbean, spread throughout the tropical world, either through trade, various means of introduction (including crop seeds contaminated with its seeds), or unintentionally through the movement of livestock (Parsons & Cuthbertson, 2001).

The earliest established records of the presence of *Ageratum conyzoides* in Southeast Asia date back to the colonial agricultural period, when it was introduced primarily through contaminated soil and crop seeds imported from Europe and Africa (Holm et al., 1997). In Malaysia, it is believed to have arrived during the rapid expansion of plantation agriculture in the early 20th century, particularly with the establishment of rubber and oil palm plantations, which provided open and disturbed habitats. Its rapid naturalisation after introduction was facilitated by its ability to produce more than 40,000 windborne seeds per plant and its capacity to adapt to various agricultural environments, including orchards, plantations, roadsides, grasslands, and highways (Sharma & Singh, 2020).

The biological and ecological characteristics of *Ageratum conyzoides* increase its likelihood of becoming invasive. One key feature is its enormous reproductive potential. A single mature plant can produce over 40,000 seeds annually, which are dispersed by wind. These seeds can remain viable in the soil seed bank for extended periods. Following the harvesting or clearing of an area, the presence of viable seeds in the soil enables *Ageratum conyzoides* to re-establish quickly. The plant produces several allelopathic compounds, such as precocene I and II, flavonoids, and benzofuran derivatives, which inhibit the germination and establishment of adjacent native plants and crops. This biochemical suppression provides *Ageratum conyzoides* with a significant competitive advantage in both natural and agricultural ecosystems (Xuan et al., 2004).

The short life cycle of *Ageratum conyzoides* (90–120 days) allows it to produce multiple generations annually in Malaysia's warm, humid climate. Additionally, the species thrives in disturbed habitats, such as construction sites, roadsides, and newly established plantations, where sunlight is abundant and competition is minimal.

C. Ecological and Economic Impacts

Ageratum conyzoides has significant ecological and economic impacts in Malaysia. Ecologically, it is an aggressive competitor against native plants, particularly within disturbed ecosystems such as forest margins, riverbanks, and areas with secondary vegetation. Its rapid growth and dense populations form thick mats that suppress the growth of native herbaceous and shrub species. The allelopathic compounds released from its leaves and roots inhibit seed germination and root development in nearby species, contributing to the loss of native plant biodiversity and alterations in community structure (Sharma et al., 2017).

The decomposition of *Ageratum* biomass may temporarily increase soil nitrogen concentrations; however, it can also disrupt the long-term balance of soil nutrients and delay the recovery of native plant communities. The presence of this species may also negatively affect pollinator interactions. Although *Ageratum* flowers attract certain bees and fly species, the dominance of this single floral resource reduces overall floral diversity and disrupts pollination networks necessary for the successful reproduction of native plant species.

Ageratum conyzoides is also regarded as a major agricultural weed affecting oil palm, rubber, pineapple, banana, coconut, and various vegetable crops due to its rapid growth and aggressive competition for soil nutrients, sunlight, and water. Farmers incur higher management costs because of the need for continuous weeding, herbicide application, and sanitation measures to prevent re-establishment (Teo et al., 2021).

In livestock grazing areas, the presence of toxic pyrrolizidine alkaloids poses risks to goats, cattle, and other grazing animals. Prolonged exposure to these toxins may cause liver damage and reduce livestock productivity (Baskaran et al., 2018). Additionally, *Ageratum conyzoides* may have adverse public health implications, as its dense growth can create favourable microhabitats for mosquito vectors such as *Aedes aegypti*, potentially increasing the risk of dengue transmission in densely populated areas lacking proper vegetation management.

D. Control and Eradication Measures

Due to its rapid growth rate and persistent soil seed banks, population control of *Ageratum conyzoides* requires a combination of methods involving continuous and sustained management. Mechanical control is most commonly practiced at the smallholder level, typically through hand-pulling or slashing before flowering occurs. In small-scale operations, hand-pulling is effective during the early growth stages; however, slashing without removing the roots allows rapid re-establishment, necessitating multiple pull-and-slash cycles.

In Malaysia, chemical control methods are widely relied upon, particularly in oil palm and rubber plantations. Herbicides such as glyphosate, 2,4-D amine, metsulfuron-methyl, and fluroxypyr have been reported to be effective in improving crop yields in affected areas. Research indicates that, when properly applied, glyphosate and metsulfuron-methyl can reduce total plant biomass by 80–90%; however, multiple applications may be necessary to fully deplete the soil seed bank (Rao & Reddy, 2014).

Ongoing research into biological control options provides insight into the feasibility of such strategies for managing *Ageratum conyzoides*. The gall-forming fly *Procecidochares utilis* has been shown to reduce seed production by inducing galls on stems and flowers, although further evaluation is required to determine its effectiveness as a biological control agent (Bhumannavar & Viraktamath, 2012). In addition, plant pathogens such as *Cercospora ageratis* have demonstrated potential as biological control agents by reducing the growth and abundance of *Ageratum conyzoides* in affected areas. However, biological control methods have not yet been widely implemented in Malaysia due to the lack of large-scale field evaluations assessing their effectiveness (Bhumannavar & Viraktamath, 2012).

Furthermore, cultural practices such as mulching, maintaining dense cover crops, and minimising soil disturbance have also been shown to effectively limit the establishment of *Ageratum conyzoides*. Therefore, Integrated Weed Management (IWM), which combines mechanical, chemical, cultural, and biological approaches, is currently regarded as the preferred strategy for effectively controlling the continued invasion of *Ageratum conyzoides.*

E. Original Field Photo

Fig. 1 *Ageratum conyzoides*

F. Invasive Species Concept Map

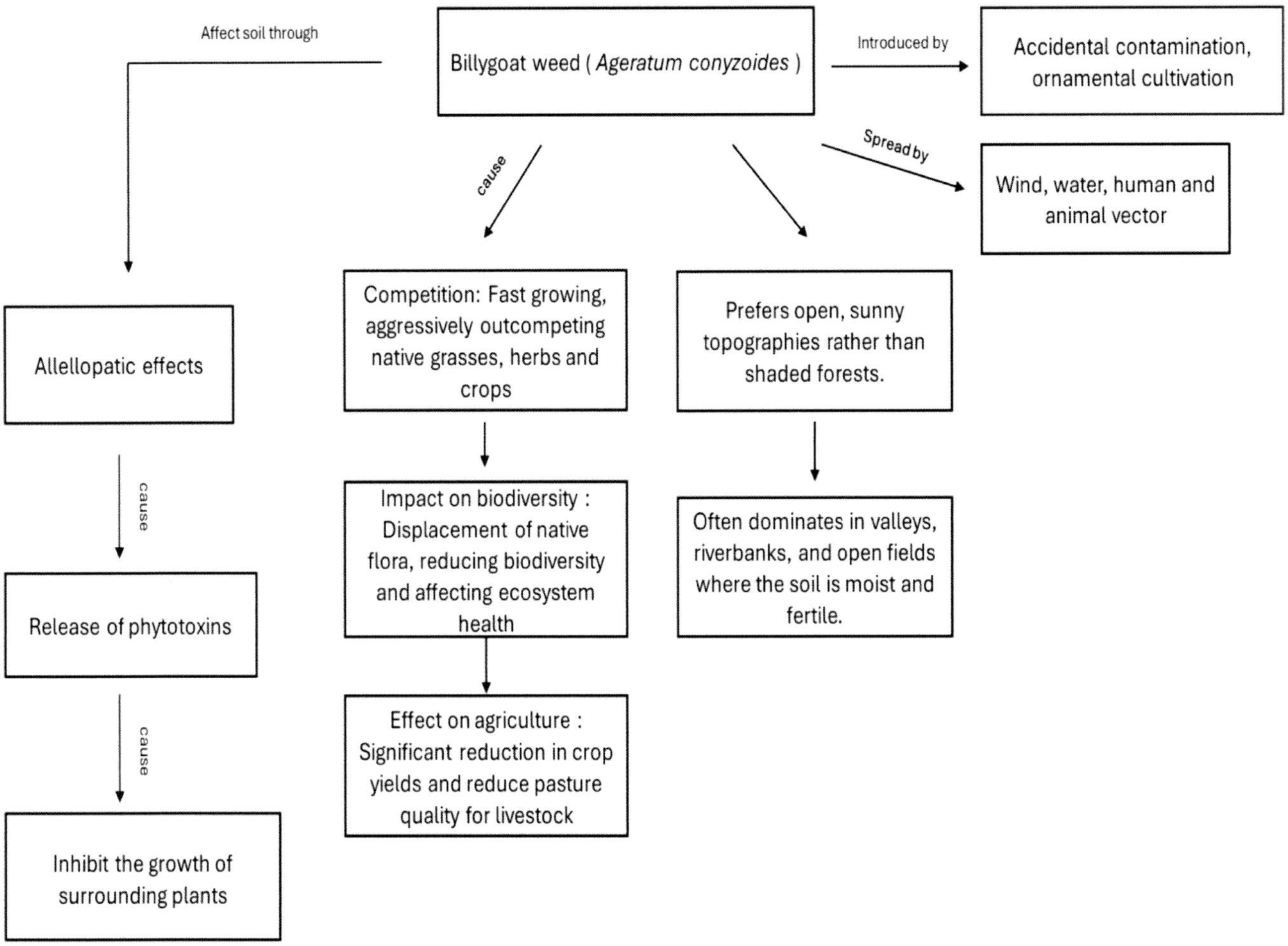

Ageratum conyzoides is an invasive species with considerable niche breadth, and its successful expansion into new habitats is underpinned by its multifaceted interactions with climate, physiography, soils, and biota. The species thrives in warm, humid tropical and subtropical conditions, where high temperatures and continuous rainfall provide optimal conditions for growth and reproduction.

In terms of physiography, *Ageratum conyzoides* favours disturbed or open habitats such as roadsides, agricultural fields, and other areas with low native plant competition. Its shallow root system allows rapid establishment across a range of site conditions, from flat lowlands to gently sloping and mountainous terrains. Soil conditions also contribute significantly to its invasion success. The species tolerates a wide variety of soil types, including relatively nutrient-poor and slightly acidic soils. It can also modify soil chemistry through the release of allelopathic compounds from its roots and decaying leaves, which reduce the germination and growth of neighbouring native plants. These chemical interactions confer a considerable competitive advantage in accessing limited resources such as light, water, and nutrients.

Biotic interactions further enhance its invasiveness. The plant's flowers attract a wide range of pollinators that facilitate seed production. Its lightweight seeds are readily dispersed by wind, water, and animals. Moreover, the presence of bitter compounds renders the plant unpalatable to many herbivores, thereby reducing grazing pressure. Collectively, these ecological traits enable *Ageratum conyzoides* to spread rapidly and form dense stands that shade and suppress native biodiversity.

The establishment of *Ageratum conyzoides* can ultimately disrupt ecological balance through altered nutrient cycling processes, degraded habitat quality, and changes to the structure and function of invaded ecosystems. Overall, the invasive behaviour of this species stems from its ability to interact effectively with both abiotic and biotic components of the environment. Its tolerance of diverse environmental conditions, coupled with its allelopathic effects and high reproductive capacity, allows it to become a dominant ecological transformer of natural habitats. Understanding these ecological interactions is essential for developing effective management strategies aimed at controlling its spread and restoring ecological balance in invaded ecosystems.

References

Baskaran, R., Perumal, K., & Sathishkumar, S. (2018). Toxic effects of *Ageratum conyzoides* on livestock: A review. *Veterinary World, 11*(2), 152–158.

Bhumannavar, B. S., & Viraktamath, C. A. (2012). Biology and control potential of *Procecidochares utilis* for managing *Ageratum conyzoides*. *Journal of Biological Control, 26*(4), 345–352.

Holm, L., Doll, J., Holm, E., Pancho, J., & Herberger, J. (1997). *World weeds: Natural histories and distribution*. Wiley.

Kumar, A., & Jhariya, M. K. (2013). Ecology and management of *Ageratum conyzoides*: A review. *Journal of Plant Science, 8*(6), 267–277.

Parsons, W. T., & Cuthbertson, E. G. (2001). *Noxious weeds of Australia* (2nd ed.). CSIRO Publishing.

Rao, V. S., & Reddy, M. S. (2014). Chemical weed management for *Ageratum conyzoides*. *Crop Protection Journal, 58*, 45–51.

Sharma, A., & Singh, R. (2020). Impact of *Ageratum conyzoides* on agricultural productivity. *Agricultural Reviews, 41*(2), 123–132.

Sharma, P., Singh, H. P., Batish, D. R., & Kohli, R. K. (2017). Allelopathic effects of *Ageratum conyzoides* on plant communities. *Annals of Agricultural Sciences, 62*(2), 211–220.

Teo, C. K., Haron, N., & Taha, R. (2021). Weed community composition and management in Malaysian plantations. *Malaysian Journal of Agriculture, 54*(1), 88–99.

Xuan, T. D., Tsuzuki, E., Terao, H., Matsuo, M., Khanh, T. D., & Hong, N. H. (2004). Allelopathic effects of *Ageratum conyzoides*. *Weed Biology and Management, 4*(4), 94–104.

Bermuda Grass (*Cynodon dactylon*): Aggressive Turfgrass Gone Wild

Auni Syazana Zahar, Fatin Nurafiqah Mohd Fikri & Nur Faiqah Zani

A. Invasive Species Name

Cynodon dactylon, commonly known as Bermuda grass, is a warm-season grass that grows in tropical and subtropical climate zones worldwide. It belongs to the Poaceae family and is widely used as a lawn grass, on golf courses, and for field sports due to its drought tolerance and rapid establishment rate. Although Bermuda grass has considerable agricultural value, it can grow aggressively and become invasive in non-native areas. Its invasive nature raises concerns regarding impacts on local ecosystems.

B. Native Range and Method of Introduction

Cynodon dactylon, commonly referred to as Bermuda grass, is native to the warm climates of Africa, southern Asia, and parts of the Mediterranean region. Over time, this species has adapted well to dry climates, particularly in sandy, low-nutrient soils typical of these regions. Due to its resilience and rapid growth rate, Bermuda grass became one of the most widely introduced turfgrass species, spreading to North America, Australia, and parts of Europe. Today, it occurs in varying abundance across these regions and has been classified as invasive in many non-native habitats (Sun et al., 2024).

Both sexual and asexual reproduction contribute to its distribution, although asexual reproduction is the primary mode of spread. While Bermuda grass produces seeds during its summer flowering stage, seed production is less reliable for rapid colonization compared to vegetative growth. Although sexually produced seeds can establish new plants, vegetative propagation through stolons (above-ground runners) and rhizomes (underground stems) plays a crucial role in its invasiveness. Under favourable weather conditions and limited disturbance, Bermuda grass can rapidly establish and become naturalised in non-native habitats.

Once established, Bermuda grass can quickly dominate, particularly in disturbed or degraded environments. Its rapid expansion and persistent growth allow it to outcompete many native species, resulting in changes to plant community composition and ecosystem structure. In agricultural systems, Bermuda grass competes with crops such as cotton, maize, and soybeans for sunlight, water, and nutrients, leading to reduced crop yields, increased management costs, and higher labour demands (Biermacher et al., 2012).

In urban environments, Bermuda grass can penetrate cracks in sidewalks, roadways, and building structures, causing infrastructure damage. Its invasive stolons and rhizomes exploit even minor openings, resulting in costly repairs and increased maintenance requirements for property owners and municipalities (Rajak et al., 2025). The combination of ecological and economic impacts makes Bermuda grass a particularly problematic invasive species. Overall, its capacity for both sexual and asexual reproduction, coupled with rapid vegetative spread through stolons and rhizomes, renders it one of the most invasive grass species. Its ability to form dense mats above ground and expand through underground rhizomes poses significant challenges for natural ecosystems, agriculture, and urban infrastructure (Callihan & Thorne, 2017).

C. Ecological and/or Economic Impacts

Cynodon dactylon can provide ecological benefits, particularly in the restoration of sodic soils. It has been reported to increase soil organic carbon, nitrogen, phosphorus, microbial biomass, soil respiration, and enzyme activity in the rhizosphere, thereby improving nutrient cycling and facilitating early ecosystem succession. Additionally, its presence can reduce soil pH, electrical conductivity, exchangeable sodium percentage (ESP), and sodium adsorption ratio (SAR) (Singh et al., 2013).

Despite these soil rehabilitation benefits, its vigorous growth can suppress native plant species and reduce overall biodiversity. When it becomes a persistent weed in croplands, its invasive nature can result in financial losses due to increased labour requirements and pesticide management costs.

D. Control and Eradication Measures

Mechanical weed control methods involve physically removing or suppressing plant growth. These include manual hand weeding and hoe weeding. Tillage may also be considered a mechanical control method when specifically implemented to directly target *Cynodon dactylon* during a crop's life cycle. However, general soil preparation practices are not classified as targeted mechanical control, even though they may indirectly affect weed populations.

Evidence suggests that hoe weeding is generally effective in reducing *Cynodon dactylon* in infested fields (Daramola et al., 2020). Studies also report high suppression levels under hand-weeding treatments, with inhibition rates exceeding 60% in many cases, indicating that hand weeding may be slightly more effective than hoe weeding under similar conditions (Akter et al., 2016). Nevertheless, mechanical control methods are labour-intensive, costly, and may inadvertently promote weed propagation through fragmentation of plant material.

Biological control methods involve the use of natural enemies such as insect herbivores, parasitoids, or pathogens to manage infestations (Omkar, 2016). In a greenhouse study, Tilley and Walker (2002) evaluated the potential of the fungus *Curvularia intermedia* as a microbial herbicide. Although plant mortality was not significantly affected, treated plants exhibited a reduction in dry weight to 38% following fungal inoculation, indicating notable susceptibility of *Cynodon dactylon*. Further research is needed to assess the effectiveness of *Curvularia intermedia*, as significant knowledge gaps remain regarding the role of insects and other microorganisms in biological control.

Chemical control methods remain widely used for large and established infestations of *Cynodon dactylon*. Effective management requires appropriate herbicide selection and timing to avoid treatment failure and resistance development. Non-selective systemic herbicides such as glyphosate are commonly applied, achieving the highest mortality rates when used during periods of active shoot and root translocation. Applications are typically repeated during active growth phases. Selective graminicides such as sethoxydim and clethodim can target *Cynodon dactylon*, although repeated applications and careful management are often necessary. In difficult-to-control sites, mixtures of glyphosate with topramezone and triclopyr have demonstrated improved suppression (Ferrell et al., 2005). However, metabolic resistance mechanisms, including cytochrome P450-mediated resistance, have been identified in *Cynodon dactylon*, potentially conferring tolerance to multiple herbicide classes.

E. Original Field Photos

Fig. 1 *Cynodon dactylon*

Figure 1 illustrates a detailed diagram of *Cynodon dactylon*, showing the overall morphology of the plant, including its slender leaves, prostrate stolons, upright culms, and inflorescence arranged in a digitate (finger-like) pattern consisting of three to seven slender spikes radiating from a single whorl at the top of the stalk. The inflorescences represent the flowering phase of *Cynodon dactylon*.

In addition to its distinctive reproductive structures, the species possesses several vegetative features that contribute to its characteristic growth form. The creeping stolons root at each node, enabling horizontal spread, rapid colonization, and the formation of dense ground-covering mats. Scaly rhizomes beneath the soil provide additional support for energy storage and vegetative propagation.

The leaves are linear and narrow, with smooth blades and a short, hairy ligule located between the leaf blade and sheath. The morphology of *Cynodon dactylon* is defined by the integration of these vegetative and reproductive traits. Erect culms arise from the stolons to support the reproductive spikes, completing the structural features that contribute to its adaptability and rapid spread.

F. Invasive Species Concept Map

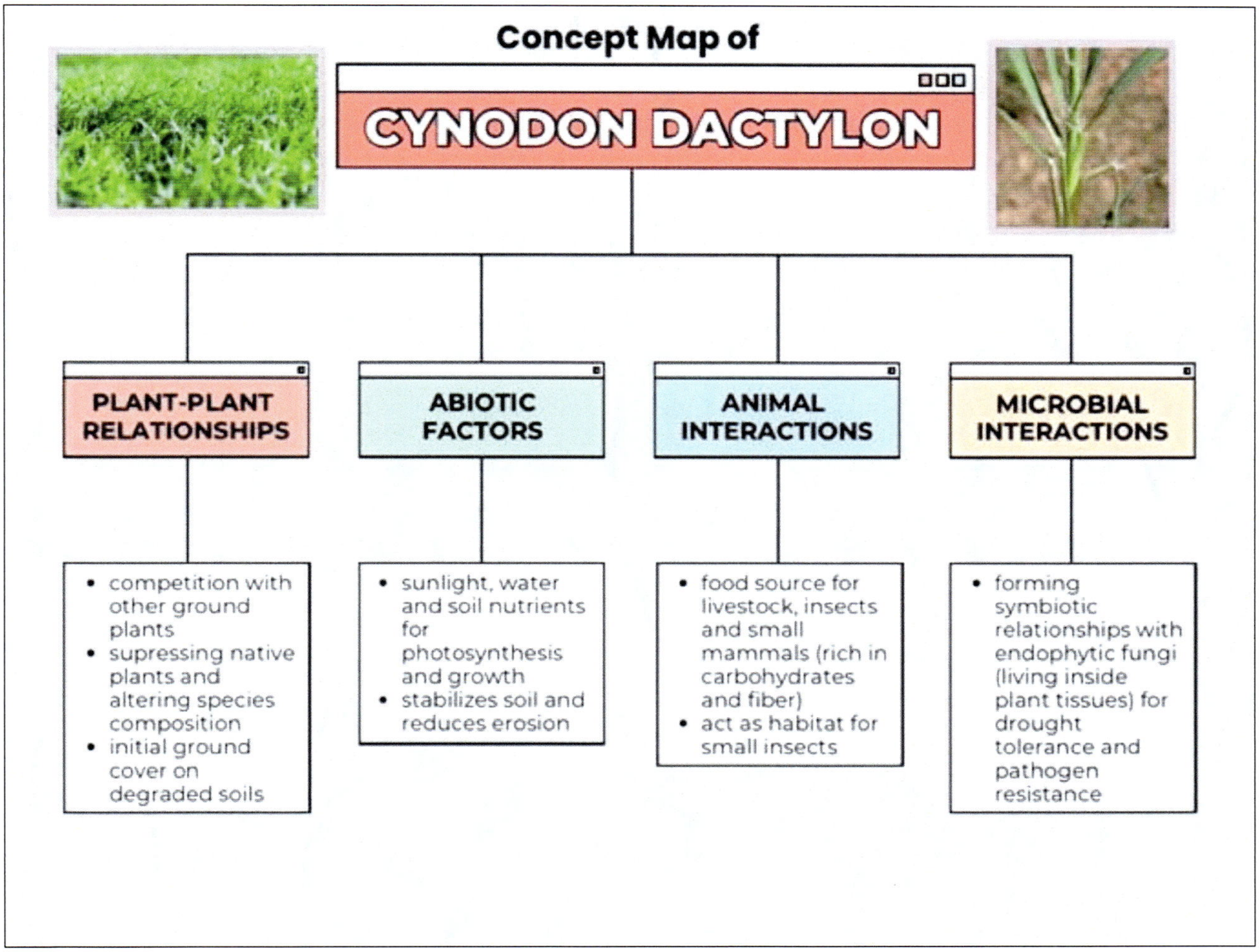

Bermuda grass competes with other ground vegetation for light, nutrients, and space. For example, it competes with *Axonopus compressus* (carpet grass). Compared to carpet grass, which has a slower growth rate, Bermuda grass spreads rapidly, particularly under warm conditions. It exhibits high drought tolerance, becoming dormant during dry periods but recovering quickly with rainfall. Generally, it is not planted alongside other grasses to prevent competition. Bermuda grass can become invasive and negatively affect biodiversity, especially in areas where native plants are fragile or slow-growing. It spreads aggressively through both seed production and creeping vegetative growth. Its dense stolons and rhizomes suppress smaller native plants, disrupting local ecosystems and altering species composition. Furthermore, it can modify soil structure and nutrient dynamics, making it more difficult for other plants to establish. During the growing season, it can spread more than two metres per month and is particularly competitive in fertile soils.

Several abiotic factors, including sunlight, water availability, and soil nutrients, strongly support Bermuda grass growth and photosynthesis, enabling it to thrive in tropical regions such as Malaysia. As a C4 photosynthetic grass, it is highly efficient in carbon fixation under high temperatures and intense light, allowing it to dominate open and disturbed landscapes. The species tolerates a wide range of soil types, including sandy loam, silty soils, and clay soils. It also demonstrates notable drought resistance due to its deep root system and rhizomatous growth habit. These characteristics enable Bermuda grass to persist in nutrient-poor or compacted soils and contribute to soil stabilisation, thereby reducing erosion in agricultural and roadside environments (Chen et al., 2015). Its rapid growth and dense ground cover can influence local microclimates by moderating surface temperature fluctuations and retaining soil moisture. Rising temperatures and increased drought associated with climate change may further enhance *Cynodon dactylon*'s competitive advantage, particularly over slower-growing native grasses.

Bermuda grass also serves as an important forage resource for livestock and as a food source for certain insects and small mammals due to its high carbohydrate and fibre content. It provides habitat for small insects and is widely used in tropical and subtropical regions for animal production. The species is favoured in intensive pasture-based livestock systems because of its tolerance to heavy grazing, as well as its resistance to disease and drought. However, grazing selectivity and pasture management practices significantly influence forage quality, nutrient cycling, and regrowth dynamics. Although Bermuda grass is recognised as a hardy forage species, its productivity and ecological balance in grazed systems depend on appropriate management.

Bermuda grass also interacts closely with microbial and fungal communities in the soil and within its tissues, which influence its health and ecological resilience. It forms symbiotic relationships with endophytic fungi that enhance drought tolerance and pathogen resistance. These fungi live within plant tissues and produce secondary metabolites that protect the grass against pathogens and abiotic stresses such as drought and salinity (Nischitha & Singh, 2025). In addition, the rhizosphere (root-associated zone) hosts various beneficial bacteria, including nitrogen-fixing species that improve nutrient uptake and soil fertility. These symbiotic relationships help maintain plant productivity even in degraded or nutrient-poor soils. Fungal associations further enhance physiological tolerance by suppressing soil-borne diseases. In return, the grass supplies carbon-based compounds derived from photosynthesis to these microorganisms. Collectively, these mutualistic interactions play an essential role in nutrient cycling and contribute to the ecological stability of Bermuda grass–dominated landscapes.

References

Akter, N., Ruhul Amin, A. K. M., Haque, N., & Masum, S. M. (2016). *Effect of sowing date and weed control method on the growth and yield of soybean. Poljoprivreda, 22*(1), 19–27. https://doi.org/10.18047/poljo.22.1.3

Biermacher, J. T., Reuter, R., Kering, M. K., Rogers, J. K., Blanton, J., Guretzky, J. A., & Butler, T. J. (2012). *Expected economic potential of substituting legumes for nitrogen in bermudagrass pastures. Crop Science, 52*(4), 1923–1930.

Callihan, R. H., & Thorne, L. (2017). Management of Bermuda grass in croplands and lawns. Agricultural Extension Service, University of Wyoming.

Chen, F., Zhang, J., Zhang, M., & Wang, J. (2015). *Effect of Cynodon dactylon community on the conservation and reinforcement of riparian shallow soil in the Three Gorges Reservoir area. Ecological Processes, 4*(1), Article 29. https://doi.org/10.1186/s13717-014-0029-2

Daramola, O. S., Adeyemi, R., Aremu, J., & Adejuyigbe, C. O. (2020). *Influence of row spacing and weed control methods on weed population dynamics in soybean (Glycine max L.). International Journal of Pest Management, 68*(1), 43–58. https://doi.org/10.1080/09670874.2020.1795300

Ferrell, J. A., Murphy, T. R., & Bridges, D. C. (2005). *Postemergence control of hybrid bermudagrass (Cynodon transvaalensis Burtt-Davy × Cynodon dactylon). Weed Technology, 19*(3), 636–639. https://doi.org/10.1614/wt-04-184r2.1

Nischitha, R., & Singh, S. K. (2025). *Endophytic fungi of Cynodon dactylon: A treasure trove of bioactive compounds. Archives of Microbiology, 207*(6), Article 207. https://doi.org/10.1007/s00203-025-04296-1

Omkar. (2016). Ecofriendly pest management for food security. Academic Press.

Rajak, P., Singh, R., Fatma, S., Nandi, A. S., & Singh, H. (2025). Economic impacts of invasive alien plant species on urban ecosystems. In Invasive alien plants in urban ecosystems: Emerging concerns and socio-ecological dimensions (pp. 201–221). Springer Nature Switzerland.

Singh, K., Pandey, V. C., & Singh, R. P. (2013). *Cynodon dactylon: An efficient perennial grass to revegetate sodic lands. Ecological Engineering, 54*, 32–38. https://doi.org/10.1016/j.ecoleng.2013.01.007

Sun, X., Fu, Q., Song, Y., Deng, X., Li, Y., Wu, K., & Fu, J. (2024). *Research progress and prospects of molecular breeding in bermudagrass (Cynodon dactylon). International Journal of Molecular Sciences, 25*(24), 13254. https://doi.org/10.3390/ijms252413254

Tilley, A. M., & Walker, H. L. (2002). *Evaluation of Curvularia intermedia (Cochliobolus intermedius) as a potential microbial herbicide for large crabgrass (Digitaria sanguinalis). Biological Control, 25*(1), 12–21. https://doi.org/10.1016/S1049-9644(02)00035-X

Touch-Me-Not (*Mimosa pudica*): Spiny Weed with Rapid Spread

Nur Ayu Natasha Abdul Muizz & Nurzatul Syahirah Sunatwi

A. Invasive Species Name

Mimosa pudica, commonly known as the sensitive plant, is a perennial herbaceous plant that belongs to the family Fabaceae. This plant typically reaches a height of about 30 cm, and its stems are covered with sharp prickles that serve as physical defense mechanisms against herbivores. These prickles make the plant difficult to handle and can cause pain or injury to grazing animals.

Its bipinnately compound leaves respond rapidly to touch. A strong touch causes the entire leaf to droop from the base of the petiole, whereas a gentle touch results in the folding of individual pairs of pinnules. According to Gupta et al. (2012), this movement is known as nyctinastic movement and is regulated by a biological clock. The leaves can also adjust their response to repeated electrical and mechanical stimulation before reopening.

Mimosa pudica produces pom-pom-shaped flowers that are pale pink or purple and composed of numerous thin, thread-like filaments. The plant reproduces by seeds, and each plant can produce more than 700 barbed seeds. These seeds can attach to human clothing, animal fur, or feathers, facilitating dispersal.

B. Native Range and Method of Introduction

This plant is native to South and Central America and is also found in parts of India. These regions provide warm temperatures and moist conditions that support rapid growth and reproduction. The successful invasiveness of this species across tropical and subtropical regions is largely due to its efficient seed dispersal mechanisms. Research indicates that the sensitive plant produces hundreds of bristly seeds that easily attach to human clothing, animal hair, and feathers, allowing long-distance dispersal. In addition, seeds can be dispersed by water during heavy rainfall or flooding events.

The seeds possess hard, glassy, or waxy seed coats. Water entry is regulated through a small opening in the seed coat, which prevents premature germination. Another pathway for the introduction of *Mimosa pudica* into new environments was through its cultivation as an ornamental plant, including bonsai forms. Its aesthetic appeal and unique leaf-closing movement have attracted plant enthusiasts.

Mimosa pudica has become invasive in many tropical and subtropical regions, including Southeast Asia, the Hawaiian Islands, Kenya, Tanzania, and Uganda. It is considered a significant pest with both ecological and economic impacts. According to Mwangi et al. (2020), it is a major weed in tropical agriculture and significantly reduces yields of important crops such as rice, maize, sugarcane, coffee, oil palm, and coconut. Its rapid growth forms dense mats that compete with crops for nutrients, light, and water.

In grazing lands, it reduces available pasture and grazing capacity by forming dense, prickly patches that restrict livestock movement. *Mimosa pudica* is also considered toxic and may cause perineal oedema and even death in livestock (Shridhar, 2017).

C. Ecological and Economic Impacts

Mimosa pudica thrives in tropical and subtropical climates. After spreading from South America to Africa and Asia, it adapted successfully to new environments. Due to its adaptability, it can displace native plant species, particularly in disturbed habitats such as roadsides. It competes with native plants for sunlight, nutrients, water, and space, potentially disrupting species with important ecological roles.

The plant forms dense, mat-like growth that suppresses other vegetation. As a nitrogen-fixing legume, it forms symbiotic relationships with nitrogen-fixing bacteria, which allow it to thrive in disturbed soils and may facilitate the establishment of other invasive weeds. According to Klonowska et al. (2012), *Mimosa pudica* can also tolerate soils contaminated with heavy metals, demonstrating strong environmental adaptability.

The spiny stems of *Mimosa pudica* cause problems for humans, livestock, and wildlife. Dense infestations restrict grazing access for goats, horses, cattle, and other animals. Animals attempting to feed or move through infested areas may suffer injuries to their legs, lips, and muzzles. Reduced grazing areas may lead to overgrazing in uninfested zones and increased soil erosion (Bhatta et al., 2023).

Mimosa pudica also competes aggressively with native plant species. The absence of natural predators in invaded areas may enhance its competitive advantage. It releases allelopathic compounds that inhibit nearby plant species. According to Huang et al. (2025), these compounds suppress root growth and alter nutrient cycling. This allows *Mimosa pudica* to dominate ecosystems by forming dense stands. Its mat-like structure also prevents the successful establishment of native seedlings, reducing plant regeneration and potentially disrupting food webs across trophic levels.

From an economic perspective, *Mimosa pudica* aggressively invades farmlands, plantations, and other managed landscapes. Farmers face reduced crop productivity and increased labour costs for weed removal. The plant's spiny stems may cause minor injuries and skin irritation among workers. Due to its rapid growth, continuous management is required, increasing maintenance costs. Dense infestations along roadsides and irrigation canals may obstruct access, block drainage systems, and raise public maintenance expenditures. Thus, *Mimosa pudica* negatively affects livelihoods and infrastructure.

D. Control and Eradication Measures

Mechanical control is effective in small or localized areas, such as home gardens. Methods include manual uprooting, slashing, mowing, and tillage. These approaches reduce flowering and plant populations. Hand-pulling can be effective but may result in minor injuries due to the plant's spines. In large areas, bulldozing combined with fire treatment may be effective. Repeated operations are necessary to deplete the soil seed bank. However, mechanical control has limitations, as the plant can regrow from root fragments remaining in the soil. Therefore, mechanical control should be integrated with other methods, such as chemical control. It is also less effective in waterlogged areas where heavy machinery cannot operate.

Biological control offers a more environmentally friendly and cost-effective approach. Natural enemies, such as insects, can be introduced to reduce infestations. Chalcodermus species (seed-feeding beetles) reduce seed production by attacking developing seeds. Other biological agents target leaves and roots. Grazing by goats and cattle may suppress new shoots. Phytopathogenic fungi such as *Cercospora pudicae* can infect *Mimosa pudica*, causing leaf spot and stem decay diseases that weaken the plant. However, the introduction of biological agents requires comprehensive environmental risk assessments to avoid unintended ecological consequences.

Chemical control is another commonly used method. Herbicides applied during early growth stages can reduce plant development. Glyphosate is often used to kill both roots and shoots. Herbicides are generally more effective when combined with mechanical methods.

E. Original Field Photos

Fig. 1 *Mimosa pudica* forming like a dense mat

Fig. 2 *Mimosa pudica* flower and flower bud

F. Invasive Species Concept Map

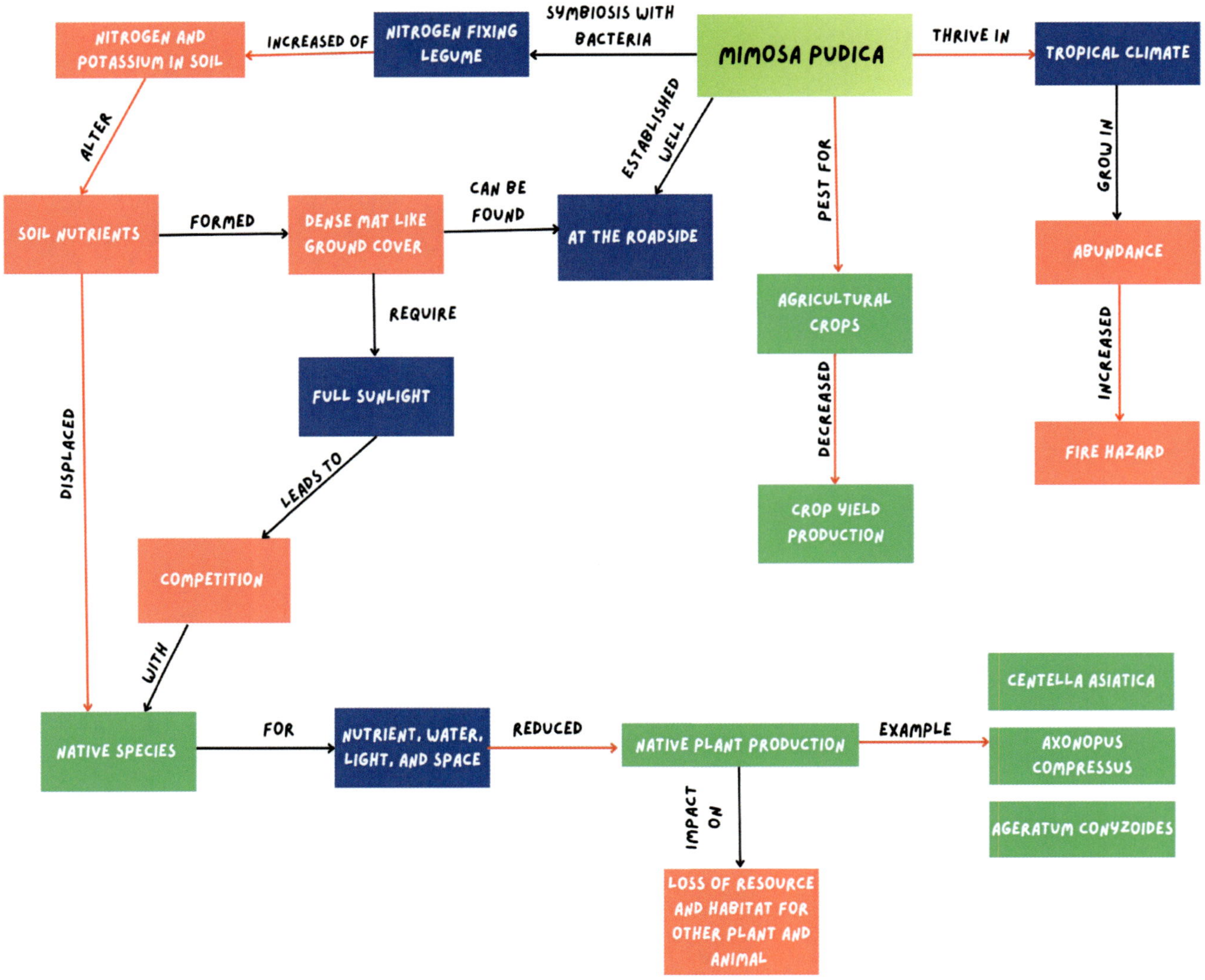

According to the concept map above, *Mimosa pudica* is classified as an invasive species in Malaysia. It thrives in tropical climates and can grow abundantly, increasing ecosystem disturbances and fire hazards. As a nitrogen-fixing legume, it alters soil nutrient dynamics, which may displace native species. It frequently forms dense roadside mats that outcompete native vegetation for nutrients, water, light, and space.

In agricultural areas, including farms and gardens, *Mimosa pudica* is a serious pest affecting crops such as rice, sugarcane, pineapple, oil palm, coconut, and coffee. Its rapid growth and dense ground cover reduce crop yields by competing for essential resources. This competition diminishes habitat availability for other plant and animal species, thereby disrupting food webs and reducing biodiversity. Native species displaced by *Mimosa pudica* in Malaysia include *Centella asiatica* (pegaga) and *Ageratum conyzoides* (billygoat weed). The overgrowth of *Mimosa pudica* suppresses these plants through competition for light and nutrients. Consequently, associated wildlife, including butterflies, small mammals, and ground-nesting birds, may also decline.

Mimosa pudica thrives in tropical climates and can grow so abundantly that it increases fire hazards, posing risks to nearby communities. When dry, the plant becomes highly flammable and contributes significantly to fire risk. In dryland agricultural systems, it is often considered a noxious weed due to its tendency to form dense ground cover even under partial shade.

References

Bhatta, S., Shrestha, B. B., & Pyšek, P. (2023). Invasive alien plants in South Asia: Impacts and management. *NeoBiota, 88*, 135–167. https://doi.org/10.3897/neobiota.88.104118

Gupta, R., Ahmad, H., Sehgal, S., & Mishra, A. (2012). *Mimosa pudica* L. (Laajvanti): An overview. *Pharmacognosy Reviews, 6*(12), 115. https://doi.org/10.4103/0973-7847.99945

Huang, H.-J., Chiu, C.-C., Lin, M.-C., Lin, C.-H., & Ye, J.-C. (2025). Allelopathic effects of common exotic plants on soil-conserving native grasses in Taiwan. *Allelopathy Journal, 65*(1), 21–34. https://doi.org/10.26651/allelo.j/2025-65-1-1528

Klonowska, A., Chaintreuil, C., Tisseyre, P., Miché, L., Melkonian, R., Ducousso, M., Laguerre, G., Brunel, B., & Moulin, L. (2012). Biodiversity of *Mimosa pudica* rhizobial symbionts (*Cupriavidus taiwanensis, Rhizobium mesoamericanum*) in New Caledonia and their adaptation to heavy metal-rich soils. *FEMS Microbiology Ecology, 81*(3), 618–635. https://doi.org/10.1111/j.1574-6941.2012.01393.x

Mwangi, H., Ita, B., Kouko, W., Maina, J. M., Momanyi, V., & Mwaja, V. (2020). *Mimosa pudica*. *PlantwisePlus Knowledge Bank: Factsheets*. https://doi.org/10.1079/pwkb.20167800554

Shridhar, N. B. (2017). Toxicity episodes of *Mimosa invisa* in cattle and buffaloes. *Toxicology International, 24*(2), 224. https://doi.org/10.22506/ti/2017/v24/i2/162430

PART III

Reflections and Lessons

Amalia Almazol, PhD

04

Across the Boundaries: Understanding Shared Invasive Traits

Across the Boundaries: Understanding Shared Invasive Traits

I. Introduction to the Chapter

This chapter examines the characteristics of an invasive species in relation to its growth and interactions with environmental conditions. It also explores the allelopathic effects of the species on the germination, growth, survival, and reproduction of other organisms in the surrounding area. It also identifies the factors that contribute to the rapid growth of invasive species wherever they are introduced, including the environmental conditions that support their establishment and persistence. These conditions include climatic, edaphic (soil-related), and biotic factors. The invasiveness of an organism is largely influenced by its capacity to adapt to diverse environmental conditions, its reproductive strategies, efficient resource acquisition, and its resilience in varying habitats.

II. Core Shared Invasive Traits

Rapid Growth

Rapid growth is one of the most common traits shared by many invasive species. This characteristic allows them to establish quickly, spread rapidly, and dominate ecosystems before native species can effectively compete. Rapid growth is largely due to the inherent biological characteristics of the species. These characteristics include the following:

Fast resource capture.

Invasive species with rapid growth rates can absorb sunlight, water, nutrients, and space more efficiently than surrounding organisms. Because they grow quickly, they can shade out smaller plants, deplete soil nutrients, and occupy the physical space required by native species. As a result, native plants often struggle to survive and compete.

a. **Early establishment advantage.**

Rapid growth also provides invasive species with a temporal advantage. They often germinate earlier in the season, grow faster during their early life stages, and reach reproductive maturity sooner. This allows them to surpass the growth of slower-growing native species and dominate habitats before native plants have the chance to fully establish.

b. **Efficient reproduction tied to growth.**

Fast growth is often associated with high reproductive output. Species that grow quickly can produce more flowers or seeds, develop more vegetative shoots or runners, and sometimes complete multiple generations within a single year. A well-known example is the water hyacinth, which can double its population in as little as two weeks. This rapid expansion forms dense mats on the surface of water bodies that block sunlight and reduce oxygen levels.

c. **Competitive suppression.**

Once a rapidly growing invasive species becomes dominant in an area, it can create environmental conditions that are unfavorable for native species. These conditions may include heavy shading, intense root competition, and altered soil conditions. Such effects can be observed in the case of Kudzu, which grows aggressively and can smother trees and other vegetation.

The following species exhibit their essential survival mechanisms of invasiveness.

The rapid growth of invasive species provides them with a significant competitive advantage over native plants. Their ability to grow quickly and adapt to a wide range of environmental conditions often makes them particularly harmful to native ecosystems. Because they can establish and spread rapidly, invasive plants frequently outcompete native species for light, water, nutrients, and space, leading to reduced biodiversity and ecosystem imbalance.

A notable example is *Celastrus orbiculatus* Thunb., commonly known as Oriental bittersweet. This species possesses several characteristics that contribute to its invasiveness. It is fire-tolerant, meaning it can survive fire disturbances and continue to grow afterward. This adaptation allows the plant to persist and spread even in ecosystems where periodic fires occur. Oriental bittersweet is also highly adaptable and can thrive under a wide range of environmental conditions. It commonly establishes in disturbed habitats such as fields, pastures, and roadsides, but it can also invade closed-canopy forests, demonstrating its tolerance for both open and shaded environments. Its vigorous root sprouting and suckering abilities enable it to quickly colonize new areas. The vine spreads aggressively, climbing and covering surrounding vegetation. As it grows, it can smother and strangle native plants, depriving them of sunlight and other essential resources. Through these mechanisms, Oriental bittersweet often outcompetes native vegetation and alters the structure of plant communities. Because of its highly competitive traits—such as shade tolerance, rapid growth, and prolific seed production—controlling and eradicating this species is extremely difficult. Despite extensive research and management efforts, an entirely effective or "perfect" eradication method for Oriental bittersweet has yet to be identified.

In the case of *Elaeagnus umbellata* Thunb., commonly known as autumn olive, the species readily establishes in disturbed, open, and edge environments. This invasive shrub typically develops multiple stems that grow into dense thickets, which can significantly alter the physical and biological dynamics of the ecosystems it invades. The formation of dense thickets allows autumn olive to intercept large amounts of sunlight, creating heavy shading beneath its canopy. As the plant reaches maturity, its canopy reduces light availability in the understory and at the ground surface. This shading can limit the growth and germination of native woody and herbaceous species, giving autumn olive a strong competitive advantage. Another factor contributing to its invasiveness is its high reproductive capacity. Autumn olive produces abundant seeds that are easily dispersed by birds and other animals, enabling the species to spread rapidly across landscapes. Autumn olive is also capable of thriving in low-fertility soils because it possesses nitrogen-fixing abilities. Through a symbiotic relationship with soil microorganisms, it can convert atmospheric nitrogen into plant-available forms such as ammonia and other nitrogen compounds. While this characteristic can be beneficial for land reclamation projects, it can also produce unintended ecological consequences. For example, autumn olive has been widely planted in reclaimed coal mining areas in Appalachia. Although its presence can increase soil nitrogen levels, it can also reduce the establishment of native plant species (Franke, Zipper, & Barney, 2018). The increase in soil nitrogen alters soil chemistry and can shift plant community composition, making the habitat less suitable for native species that are adapted to nutrient-poor soils. As a result, while autumn olive's nitrogen-fixing ability allows it to thrive in degraded environments, it can also disrupt native plant communities and ecosystem processes, further reinforcing its status as an invasive species.

The characteristics of *Alliaria petiolata* (M. Bieb.) Cavara & Grande (Garlic mustard) pose a significant threat to the native biodiversity and hardwood forests of West Virginia. This invasive species often forms dense, monotypic stands that aggressively outcompete and shade native vegetation. One of its key advantages is its extended growing season. Garlic mustard emerges earlier in the spring and remains green later into the fall than many native plants. This longer growth period allows it to capture light, nutrients, and space before native species begin active growth. As a result, garlic mustard can displace many native spring ephemeral wildflowers, including species such as Trillium, Spring beauty, and Toothwort. These plants depend on early-season sunlight before the forest canopy closes, but the presence of dense garlic mustard populations can significantly reduce their growth and survival. Overall, the most significant ecological impact of garlic mustard is its ability to compete with and displace native plant species, ultimately reducing biodiversity and altering the composition of forest understory communities.

The *Berberis thunbergii DC.* commonly known as Japanese barberry, is highly adaptable to a wide range of environmental conditions. It is tolerant of both full sunlight and deep shade, allowing it to thrive in many different habitats and light environments. This adaptability enables the shrub to establish in closed-canopy forests, open woodlands, wetlands, and field edges, making it particularly successful across diverse landscapes. Japanese barberry often forms dense stands that grow rapidly and shade out surrounding vegetation. As these thickets expand, they can suppress the growth of native plants and gradually dominate the understory layer. Over time, these dense populations may develop into near-monocultures, replacing the diverse native plant communities with large concentrations of this invasive shrub. Seed dispersal also contributes significantly to its spread. Birds—including wild turkeys and other game birds—consume the plant's bright red berries and disperse the seeds across wide areas. This allows Japanese barberry to colonize new habitats far from the original parent plant. Another factor contributing to its invasiveness is the lack of natural herbivores in North America that regularly browse on the shrub. Without significant browsing pressure from mammals, the plant can grow and spread with relatively little biological control. Additionally, Japanese barberry reproduces through both sexual reproduction (via seeds) and asexual reproduction (such as vegetative growth). Each seed has the potential to develop into a new shrub, allowing populations to expand and spread into surrounding ecosystems. As a result, this combination of adaptability, rapid growth, and effective dispersal mechanisms enables Japanese barberry to become a persistent and dominant invasive species in many environments.

The *Lonicera japonica* Thunb., commonly known as Japanese honeysuckle, is a highly invasive vine characterized by its extensive vegetative growth and rapid spread. The plant forms underground rhizomes, layered stems, and lateral roots, and it can also reproduce from plant fragments. These growth strategies allow it to expand rapidly in patches and establish local dominance within invaded areas. Originally introduced through intentional planting, Japanese honeysuckle quickly escaped cultivation due to its vigorous growth and effective reproduction. It spreads through both vegetative propagation and seed dispersal by birds and mammals, which has enabled the species to invade surrounding wildlands. Japanese honeysuckle is considered an aggressive invasive species that commonly colonizes forest edges, open woodlands, riparian zones, roadsides, early successional fields, and some wetlands. In these habitats, the vine forms dense mats along the ground and climbs over shrubs and small trees. As it grows, it can smother surrounding vegetation, significantly altering the structure and composition of native plant communities. The ecological impacts of Japanese honeysuckle are particularly evident in its interception of light and competition with native species. Its vining and climbing growth habit allows it to shade the understory, reducing the amount of sunlight available to native plants. This shading suppresses the regeneration of tree seedlings and herbaceous species, ultimately limiting the recovery and diversity of native vegetation.

The *Pueraria montana* var. *lobata* (Willd.) Maesen & S. M. Almeida ex Sanjappa & Predeep, commonly known as kudzu, is an exceptionally damaging invasive species in the Southeastern United States due to its rapid growth and vining habit, which allow it to smother native vegetation and block sunlight. Its aggressive growth makes it one of the most recognizable and culturally significant invasive plants in the region. Kudzu has serious ecological impacts, primarily by forming dense monocultures that suppress native plant species, reduce biodiversity, and alter ecosystem functions. Covering trees, shrubs, and the forest floor, it limits light availability, which can impede the regeneration of native species and reduce carbon sequestration over time. In addition to ecological concerns, kudzu poses significant economic threats. It can carry the pathogen responsible for soybean rust, which survives over the winter on kudzu. Kudzu also threatens the timber industry (Harron, Joshi, Edgar, Paudel, & Adhikari, 2020). Overall, kudzu's ability to shade out native vegetation and timber crops makes it both an ecological and economic concern, highlighting the dual consequences of invasive species that grow rapidly and dominate landscapes.

Rosa multiflora Thunb. (Multiflora rose) is a deciduous shrub capable of thriving in a wide range of habitats. It frequently forms dense thickets along roadsides, forest edges, open fields, and other disturbed sites. The coloration of its stems varies seasonally and with age, reflecting its dynamic growth patterns. Multiflora rose's rapid growth has widespread ecological consequences. It can outcompete native plants by choking out roots, monopolizing soil nutrients and moisture, and forming dense thickets that limit access for other species. Displacing native flowering plants, it can reduce floral diversity and disrupt the foraging behavior of pollinators. The dense thickets also influence wildlife movement. Animals such as deer may be forced to alter their routes, avoiding areas dominated by multiflora rose, which can affect feeding patterns and ecosystem dynamics. In forests, the shrub's dominance in the understory can suppress the regeneration of native tree seedlings, altering natural forest succession and long-term forest development. Multiflora rose poses additional ecological risks. It can hybridize with native rose species, potentially threatening their genetic integrity. Its dense, high-biomass growth also alters fire dynamics, increasing fuel loads and creating hazards during wildfires. Overall, multiflora rose's combination of rapid growth, dense thicket formation, and aggressive competition allows it to significantly alter ecosystem structure, biodiversity, and natural processes.

The *Ailanthus altissima* (Mill.) Swingle, commonly known as Tree of Heaven, is a deciduous tree that has become highly invasive across temperate regions of North America and Europe. Its widespread invasiveness, combined with its striking appearance, has earned it the nickname the "alien tree." Tree of Heaven can grow to heights exceeding 80 feet and is capable of thriving in a wide range of environmental conditions. Its rapid growth rate and resistance to disease and insect pests have contributed both to its popularity in cultivation and to its success as an invasive species. The ecological consequences of its spread are significant. Tree of Heaven can outcompete native vegetation, alter forest composition, and disrupt ecosystem processes. Its proliferation is facilitated by a combination of biological traits—such as fast growth, prolific seed production, and tolerance of poor soils—and external factors, including human-mediated disturbance and urban planting. Controlling this species presents considerable challenges due to its resilience and ability to regenerate from root suckers. Understanding its ecological impacts, spread mechanisms, and management difficulties is critical for mitigating the effects of this invasive tree on native ecosystems.

The *Pontederia crassipes* Mart., commonly known as water hyacinth, is a highly prolific invasive species with both sexual and asexual reproductive capabilities. Through sexual reproduction, a single flower can produce up to 300 seeds, which remain viable for up to 20 years. In addition, a single plant can generate more than 3,000 daughter plants within two months through vegetative propagation. This extraordinary reproductive capacity enables water hyacinth to spread rapidly, forming dense mats that drastically reduce dissolved oxygen levels and threaten aquatic biodiversity. These dense mats also obstruct waterways, interfering with boating, fishing, and other human activities. Ecologically, the mats block sunlight, displacing native aquatic vegetation and altering trophic dynamics. This leads to habitat degradation and a decline in ecosystem health. A key trait driving water hyacinth's invasiveness is its high biomass accumulation and clonal propagation. The species primarily reproduces vegetatively via stolons, allowing exponential population growth under favorable conditions. Empirical studies demonstrate that in nutrient-enriched waters, water hyacinth can rapidly increase both surface coverage and biomass, often outcompeting native macrophytes (Gopal, 1987; Wilson et al., 2005). Its growth strategy mirrors that of many invasive terrestrial plants, which exploit canopy gaps or edge environments to monopolize resources through rapid lateral or vertical expansion. In aquatic systems, dense hyacinth mats restrict light penetration, suppress submerged vegetation, and reduce oxygen concentrations, fundamentally altering ecosystem processes (Villamagna & Murphy, 2010). Water hyacinth also thrives in disturbed, nutrient-rich environments, such as eutrophic water bodies impacted by agricultural runoff, sewage, and urban development. These conditions are analogous to the human-modified landscapes that facilitate terrestrial plant invasions. Its physiological tolerance to fluctuating water levels, poor water quality, and chemical stressors allows it to persist where native species experience reduced fitness, further accelerating invasion. In summary, water hyacinth's combination of rapid sexual and asexual reproduction, high biomass accumulation, environmental tolerance, and association with nutrient-enriched habitats makes it one of the most aggressive and ecologically disruptive invasive aquatic plants.

The *Lantana camara* L. (Common lantana) is a highly adaptable plant that thrives in tropical and subtropical climates. Within its native and introduced ranges, it can establish across a variety of ecosystems, often displacing indigenous vegetation, altering soil chemistry, and disrupting forest regeneration. The species reproduces prolifically, producing large quantities of seeds that remain viable in the soil for extended periods. This creates a persistent seed bank, enabling rapid re-establishment and colonization of disturbed habitats such as forest edges, grasslands, abandoned farms, and roadsides. Its fruits are dispersed effectively by birds and other animals, further facilitating its spread (Kusumoto et al., 2023). Once established, common lantana forms dense thickets that grow rapidly, allowing it to aggressively outcompete seedlings, herbs, and understory plants. This suppression of native vegetation alters community composition and reduces biodiversity. Due to its persistence and regenerative capabilities, eradication requires a comprehensive management strategy that integrates mechanical removal, chemical treatment, and biological control. The combination of rapid growth, prolific seed production, and effective dispersal makes common lantana a highly invasive and ecologically disruptive species across affected landscapes.

Mikania micrantha Kunth (Bitter vine) is a highly invasive species distinguished by its rapid growth, high seed production, vegetative reproduction, and broad environmental tolerance. These traits allow it to spread quickly and dominate landscapes, forming dense mats that smother native vegetation, blocking sunlight and limiting access to essential nutrients. This aggressive growth profoundly alters ecosystem balance and affects ecological, economic, and environmental components of the habitats it invades. Bitter vine preferentially flowers in open, sunny areas, while shaded regions are less favorable for its reproduction. The species reproduces both sexually via wind-dispersed seeds and vegetatively through stem fragments. Its seeds are equipped with pappi that can attach to human clothing, equipment, or animal fur, making humans and animals effective agents of dispersal. The main stem of *Mikania micrantha* is also capable of sprouting upon contact with soil and can survive drought for extended periods, resuming growth under favorable conditions. These combined traits—rapid vegetative growth, persistent stems, efficient seed dispersal, and environmental resilience—make bitter vine a highly adaptable and ecologically disruptive invasive species.

Swietenia macrophylla King, commonly known as big-leaf mahogany, is recognized for its role as a pioneer species in degraded agricultural landscapes, where it can outcompete other plantation species. In the Philippines, it has been valued for its wind firmness and resistance to cyclone damage, making it an attractive choice for silvicultural and timber production purposes. Despite these advantages, *Swietenia macrophylla* is now considered one of the most ecologically threatening invasive tree species in the country. Coracero (2023) documented its spread in the buffer zones of the Mt. Banahaw–San Cristobal Protected Landscape in Nagcarlan, Laguna side, highlighting its ability to expand beyond plantation areas into natural ecosystems. The species' invasiveness in the Philippines is driven by several key traits: long-distance wind dispersal, rapid growth, and strong dominance in disturbed forest areas. These characteristics allow big-leaf mahogany to establish and proliferate quickly, outcompeting native plants and impeding the regeneration of indigenous species, thereby altering forest composition and ecosystem dynamics.

Leucaena leucocephala (Lam.) de Wit (Ipil-ipil) has significant ecological impacts in invaded habitats, largely due to its rapid growth and ability to outcompete the native species in the area. It forms dense monospecific stands that suppress neighboring woody and herbaceous vegetation through both competition for light and allelopathic effects; its leaf litter has been shown to inhibit the germination and growth of a variety of species (Sharma et al., 2022). If left unmanaged during the growing season, stand density can increase greatly due to seed drop. Seed can be further disseminated by passage through the grazing animal or by water movement. Thickening up of hedgerows may occur where delayed grazing allows seed set and seedling recruitment.

Leucaena leucocephala (Lam.) de Wit, commonly known as ipil-ipil, has significant ecological impacts in the habitats it invades. Its rapid growth and competitive dominance allow it to outcompete native vegetation, often forming dense monospecific stands. These stands suppress surrounding woody and herbaceous plants through both competition for light and allelopathic effects, as their leaf litter has been shown to inhibit the germination and growth of a variety of species (Sharma et al., 2022). If left unmanaged during the growing season, stand density can increase substantially due to prolific seed production. Seeds are easily dispersed through multiple mechanisms, including passage through grazing animals or transport via water movement. In areas where grazing is delayed, hedgerows may thicken, allowing seed set and seedling recruitment to further enhance the spread of ipil-ipil. Overall, its combination of rapid growth, allelopathic influence, high seed production, and effective dispersal makes Leucaena leucocephala a highly invasive species capable of altering plant community structure and reducing native biodiversity.

Thunbergia grandiflora (Roxb. ex Rottler) Roxb. (Sky flower) is an aggressive invasive species that reproduces both sexually through seeds and asexually via cuttings, contributing to its rapid growth and widespread colonization. Sky flower continuously invades and outcompetes native vegetation, smothering host trees and understory plants. Its dense growth suppresses the germination and development of native seedlings, further reducing the regeneration of indigenous species. The vine has the capacity to dominate entire acres of vegetation, creating monocultures that do not support native wildlife, which intensifies its ecological impact. These characteristics make Thunbergia grandiflora a particularly serious threat to native plant communities, especially in lowland tropical rainforests, where its invasion can lead to substantial biodiversity loss and disruption of natural ecosystem processes.

Mimosa invisa Martius ex Colla (Makahiyang lalaki) is a highly invasive species capable of colonizing grasslands, fields, plantations, and forest margins. It often forms dense, monocultural stands, which enable it to dominate large areas and spread rapidly. The aggressive growth of *Mimosa invisa* poses a significant threat to native ecosystems and biodiversity, as it outcompetes local plant species for resources. Studies measuring vegetative shoot density per unit area indicate that its invasiveness can substantially reduce populations of native flora, leading to a loss of plant diversity and disruption of ecosystem structure. Its combination of rapid spread, dense stand formation, and competitive dominance makes Makahiyang lalaki a major ecological concern in invaded habitats.

Imperata cylindrica (L.) P. Beauv. commonly known as cogon grass, is a highly invasive species that can exert landscape-scale impacts on ecosystem structure and function. The species forms extensive, continuous stands, often dominated almost entirely by cogon grass. These dense populations outcompete native vegetation for light, water, and nutrients, resulting in reduced plant diversity and disrupted ecosystem stability. Cogon grass also significantly influences fire dynamics. Its leaves and stems contain high oil content and accumulate large amounts of dry biomass, making the plant highly flammable, particularly during dry seasons. As a result, areas dominated by cogon grass often experience increased frequency, intensity, and severity of wildfires. These fires can kill many native plant species, while cogon grass survives due to its protected underground rhizomes, which allow it to rapidly resprout and quickly reclaim the burned area. The species thrives best in sunny and moist environments, where it can spread aggressively and form dense mats. These mats suppress surrounding vegetation and may limit moisture infiltration to competing plants, further enhancing cogon grass's competitive advantage. Overall, the combination of dense stand formation, strong competitive ability, and fire-adapted survival strategies enables cogon grass to dominate invaded landscapes and significantly alter ecological processes.

Allelopathy

Allelopathy refers to the biological phenomenon in which plants influence neighboring plants or microorganisms through the release of chemical compounds (Schandry & Becker, 2020). These chemical interactions can affect the growth, survival, and reproduction of surrounding organisms within an ecosystem. According to Cheng and Cheng (2015), allelopathy describes the direct or indirect effects of one plant on another, including interactions mediated through microorganisms. It is essentially a form of chemical interference among plants, where compounds released into the environment can either stimulate or inhibit the growth of nearby vegetation (Fernandez et al., 2016). Bahadur et al. (2015) further defined allelopathy as the injurious effect of one organism on another through the production of phytotoxic compounds, which influence the development and growth of neighboring plants. Expanding upon earlier definitions, the International Allelopathy Society broadened the concept in 1996 to include any process involving secondary metabolites produced by plants, microorganisms, bacteria, and fungi that can either stimulate or inhibit growth and development within agricultural and natural biological systems (Schandry & Becker, 2020). Because of these chemical interactions, allelopathy can play a significant role in plant invasiveness. Some invasive species release compounds that suppress the germination and growth of neighboring plants, allowing them to dominate habitats and outcompete native vegetation. As a result, allelopathic mechanisms are often considered an important factor contributing to the competitive success of invasive plant species in many ecosystems.

Allelopathy can be both an evolutionary advantage and a contributor to ecological imbalance. Allelopathy is an evolutionary advantage because it is a competitive strategy shaped by natural selection. Plants can release inhibitory chemicals into the soil to gain their advantages, such as reducing competition for light, water, and nutrients; preventing nearby seedlings from establishing; and altering microbial communities in the soil, favoring their establishment, growth, and survival. On the other hand, allelopathy is a driver of ecological imbalance when a species enters an ecosystem where other organisms are being suppressed by its chemical defenses. Usually, native species are not resistant to chemicals. Soil microbes may break down the compounds effectively, and competitive suppression becomes extreme, as in the case of Garlic mustard in North American forests. Its chemicals disrupt soil fungi that native trees depend on, giving it a major advantage over native vegetation.

The following invasive species have their own mechanisms of affecting other species through allelopathy:

The presence of *Celastrus orbiculatus* Thunb. (Oriental bittersweet) has been associated with changes in soil chemistry and nutrient dynamics. Studies have shown that infestations of Oriental bittersweet can alter soil pH and the concentrations of essential nutrients, including calcium, potassium, and magnesium. In addition, the species has been linked to increased nitrogen mineralization and accelerated litter decomposition rates, which can further modify soil processes and influence plant community structure (Wooten, 2013).

In natural or reclaimed landscapes, vegetation succession typically begins with the colonization of herbaceous plants, shrubs, and eventually woody tree species. However, the aggressive establishment of *Elaeagnus umbellata* (autumn olive) in open sites can disrupt this natural successional process. Because autumn olive establishes quickly and forms a persistent shrub layer, it can alter or even prevent the development of a tree-dominated forest system. One key factor contributing to this disruption is the plant's nitrogen-fixing ability, which allows it to increase nitrogen availability in otherwise nutrient-poor soils. While this trait may improve soil fertility, it can also favor fast-growing shrubs and opportunistic plant species over slower-growing native tree seedlings that are adapted to low-nutrient conditions. Evidence from reclaimed mine sites supports this effect. Franke, Zipper, and Barney (2018) found that the survival rate of native tree seedlings decreased significantly in areas where autumn olive was present. When removal treatments were implemented, tree seedling establishment improved over two growing seasons. These findings suggest that the presence of autumn olive can delay or prevent the natural progression of succession toward a hardwood forest canopy in reclaimed landscapes.

Alliaria petiolata (Garlic mustard) exhibits strong allelopathic properties, releasing toxic chemical compounds, known as allelochemicals, into the surrounding soil. These compounds inhibit the growth of competing native plants and disrupt important ecological relationships. In particular, garlic mustard interferes with the symbiotic relationship between native tree species and mycorrhizal fungi, which many forest plants rely on for nutrient uptake. This disruption can significantly suppress the growth and survival of native tree seedlings, including species such as Oak. Another major ecological impact of garlic mustard involves its interaction with the West Virginia white butterfly. Garlic mustard acts as a "toxic decoy" for this species. Because the plant belongs to the Brassicaceae family, the same family as the butterfly's native host plants, such as Toothwort, female butterflies are attracted to it and deposit their eggs on its leaves. However, the larvae cannot survive on garlic mustard, making the plant a "toxic trap." This deceptive interaction has contributed to the decline and local extirpation of the West Virginia white butterfly in some regions. Through both allelopathic soil effects and harmful ecological interactions with native wildlife, garlic mustard significantly disrupts forest ecosystems and threatens native biodiversity.

Berberis thunbergii, commonly known as Japanese barberry, can significantly alter plant communities and soil processes in areas where it becomes well established. In heavily invaded sites, the shrub displaces many native herbaceous and woody species, reducing plant diversity and altering the structure of forest understories. Large infestations of Japanese barberry can also modify soil chemistry. Its leaf litter and root system contribute to changes in soil conditions, often increasing soil pH and making the soil more basic. These altered soil conditions can create an environment that is more favorable for the continued establishment and spread of Japanese barberry, reinforcing its dominance in invaded areas. In addition, Japanese barberry has been shown to affect forest soil nutrient cycling through its interactions with invasive earthworms. These earthworms accelerate the decomposition of leaf litter and modify soil structure, which can further alter nutrient availability and ecosystem processes. Together, these changes can contribute to long-term shifts in forest soil dynamics and plant community composition.

Lonicera japonica (Japanese Honeysuckle) is most problematic across the eastern United States, where environmental conditions favor its rapid growth and spread. In colder northern regions, its growth tends to be limited or seasonal, as low temperatures restrict its activity during part of the year. In contrast, in warmer areas, Japanese honeysuckle may remain semi-evergreen, allowing it to grow for longer periods and become more aggressive in its spread and ecological impact.

Ailanthus altissima, commonly known as the tree of heaven, has significant ecological impacts in areas where it becomes established. The species reproduces rapidly and forms a dense canopy that blocks sunlight from reaching the forest floor, which suppresses the growth of other plant species. In addition, it spreads through aggressive root suckers, allowing it to expand quickly while competing with surrounding vegetation for soil nutrients. The tree is also allelopathic, releasing a toxic compound called ailanthone into the soil. This chemical inhibits the germination and growth of neighboring plants, giving the species a competitive advantage in colonizing disturbed environments (Sladonja et al., 2015). Because of these chemical secretions, the tree of heaven has sometimes been referred to as "stinking sumac," as its compounds produce a strong odor often described as similar to rotting flesh or burned peanut butter.

Swietenia macrophylla (Big-leaf Mahogany) has also been reported to exhibit allelopathic effects in areas where it becomes established. Studies suggest that the species may release chemical compounds from its leaves and decomposing litter that can inhibit the germination and growth of nearby plant species. In addition, the accumulation and decomposition of its leaf litter may alter soil nutrient composition, potentially creating conditions that are less favorable for the establishment and regeneration of native vegetation. These effects may contribute to the species' ability to dominate disturbed habitats and limit the recovery of native plant communities.

Leucaena leucocephala, commonly known as ipil-ipil, exhibits strong allelopathic effects that influence plant community dynamics and ecosystem processes. Studies have shown that its leaf litter, mulch, and seed residues can suppress the germination and growth of numerous plant species. For example, Leaf litter mixed with soil inhibited the germination and growth of test woody plant species (Ahmed et al., 2008). Soil containing decomposing leaves increased mortality in *Alnus formosana, Acacia confusa, Liquidambar formosana, Casuarina glauca, and Mimosa pudica* (Chou & Kuo, 1986). Aqueous extracts of its litter suppressed germination and radicle growth in *Lolium multiflorum, Ageratum conyzoides, Tridax procumbens, and Emilia sonchifolia* (Chou & Kuo, 1986; Ishak et al., 2026). Leaf mulch applied to soil inhibited *Vigna unguiculata* germination and root nodulation (Kamara et al., 1999). These observations suggest that *Leucaena leucocephala* contains allelochemicals in its leaves and seeds, some of which are released into the soil during decomposition (Kato-Noguchi & Kurniadie, 2022). Its high nitrogen-fixing capacity further alters soil nutrient dynamics, increasing nitrogen availability and potentially favoring the establishment of other invasive species, thereby disrupting native community structure (Sun et al., 2025). The species' reproductive and life-history traits, including abundant seed production, vigorous resprouting, and production of defensive compounds such as mimosine, enhance its invasiveness. In regenerating forest fragments, dominance by ipil-ipil can reduce species richness and functional diversity, particularly among animal-dispersed species, which may compromise long-term ecosystem recovery (Santos et al., 2023). Its tolerance to disturbance allows it to resprout after fire or cutting, enabling it to maintain dominance and reshape vegetation patterns, succession trajectories, and overall community assembly (Sharma et al., 2022). Chemically, the leaves and seeds contain lipids, crude protein, carbohydrates, tannins, oxalic acid, and the toxic non-protein amino acid mimosine, while the seed kernel has an oil content of 17–20% (Gutteridge et al., 1998; Hossain et al., 1998). These biochemical traits contribute to both its allelopathic effects and ecological resilience, making Leucaena leucocephala a highly disruptive invasive species.

Mimosa invisa, commonly known as makahiyang lalaki, is commonly found in damp waste areas, along roadways, and in disturbed landscapes. The species poses a significant threat to croplands, pastures, sugarcane and pineapple fields, as well as coconut, tea, and rubber plantations, where dense infestations can severely impact agricultural productivity. Its thick growth inhibits the germination, growth, reproduction, and regeneration of native plant species, often forming dense monocultures that displace indigenous vegetation. All parts of the plant are toxic to herbivores, producing the compound mimosine, a non-protein amino acid. Mimosine can damage the endothelium, cause liver and heart necrosis, and induce anemia in cattle, making Mimosa invisa both an ecological and economic concern in invaded areas.

Pontederia crassipes, commonly known as water hyacinth, extends its invasive success beyond physical dominance by employing chemical interference strategies. Studies indicate that the species releases secondary metabolites into surrounding waters, which negatively impact the growth and physiological performance of co-occurring aquatic plants and phytoplankton (Malik, 2007). This allelopathic capacity is analogous to mechanisms observed in invasive terrestrial plants, which can alter soil chemistry to inhibit the establishment and growth of native species. In both aquatic and terrestrial ecosystems, allelopathy serves as a non-resource-based competitive strategy, enabling invasive species to restructure community composition and reinforce their dominance, independent of direct competition for light, nutrients, or space.

Imperata cylindrica, commonly known as cogon grass, exhibits allelopathic properties that enhance its competitive dominance in invaded habitats. The species releases chemical compounds into the soil that inhibit the germination and growth of neighboring plants, allowing it to monopolize resources and reduce biodiversity. Its extensive rhizome network, combined with these allelopathic effects, facilitates rapid spread and long-term persistence. Cogon grass thrives in disturbed, nutrient-poor soils and actively modifies soil chemistry by altering levels of silica, organic matter, and other nutrients. These changes can shift soil pH and nutrient cycling, which in turn affects microbial and fungal communities, disrupting essential symbiotic relationships. Through this combination of chemical interference, vegetative growth, and soil modification, Imperata cylindrica significantly transforms ecosystem structure, function, and species composition.

Lantana camara, commonly known as common lantana, is characterized by rapid growth, high reproductive capacity, and allelopathic effects on surrounding vegetation. The species releases chemical compounds that can suppress the germination and establishment of native plants, limiting their recruitment and growth. As a result, lantana invasions often lead to reduced biodiversity and altered habitat functions in affected ecosystems (MDPI Plants, 2021). *L. camara* releases allelopathic chemicals from its leaves and litter that inhibit the germination and growth of understory plants. This suppression of native seedlings disrupts forest regeneration and alters plant community composition. Over time, the cumulative impact of lantana invasions can lead to habitat conversion, transforming areas of diverse native vegetation into landscapes structurally dominated by this invasive species. Such changes reduce the availability of ecological niches for native flora, contributing to long-term declines in biodiversity and ecosystem complexity.

Disturbed Habitats

Invasive species are often a symptom of human disturbance, though the relationship is nuanced. In many ecosystems, invasions are strongly linked to human activities that disrupt natural barriers and ecological balance. Humans move species beyond their natural ranges. Through global trade, humans move organisms intentionally or accidentally through shipping, agriculture, pet and aquarium trades, and the global transportation network. When the species moved to their new environment, they may become invasive because of the absence of natural predators/enemies, parasites, or competitors, hence they spread rapidly. Moreover, these species have an efficient root system that can utilize the available water, nutrients, sunlight, and space, boosting their rapid growth and colonization, which in turn suppresses the native species in the area.

Disturbance makes an ecosystem easier to invade as it often weakens resistance, giving invaders opportunities to establish. These disturbances include deforestation, urbanization, agriculture, pollution, climate change, and overgrazing. This disturbance alters the species composition, soil and water conditions, and creates empty ecological niches. Disturbed habitat tends to favor fast-growing and adaptable species, which makes them invasive. Once established, these species can create feedback loops that make an ecosystem even more disturbed because they amplify disturbance, as in the case of *Imperata cylindrica,* which is prone to wild fire increases fire frequency, making it harder for native species to recover. However, not all invasive species require disturbance. Some invasive species can invade relatively intact ecosystems if they have strong competitive advantages. Hence, human disturbance introduces species and weakens ecosystems that make invasions possible, but invasive species themselves become drivers of further ecological change.

A disturbed environment is favorable for the colonization, growth, and survival of the invasive species. Some invasive species are being introduced in the area as edible plants for medicinal and culinary purposes. Other species are being planted for landscape and beautification of the surroundings, while other species are used as reforestation planting materials to enrich the disturbed environment. Since these species are usually pioneer and tolerant to fire and acidic soil, they can survive and utilize the minimal nutrients present in the area and release allelopathic chemicals that suppress the native vegetation. Further, the seeds of these invasive species can be dispersed freely by biotic and abiotic agents of dispersal, such as by themselves, animals including humans, wind, and water. The disturbed environment could enhance the germination

Effective control of Garlic Mustard demands a long-term commitment, often spanning 5–10 years, primarily due to its persistent seed bank, where seeds can remain viable in the soil for at least five years.

III. Cross-Boundary Patterns

Note that not all traits are universal predictors of invasiveness. Traits like rapid growth, high reproduction, broad tolerance, and strong dispersal are common among invasive species, but they are not universal predictors of invasiveness. Ecologists generally agree that no single trait—or even small set of traits—reliably predicts which species will become invasive in every ecosystem. Invasions depend on a combination of species traits, environmental conditions, and human activity.

1. Many non-invasive species have the same traits

Traits often associated with invasiveness are actually very common among many organisms, including species that never become invasive. Many fast-growing plants remain harmless in their native ranges. Some species with high reproductive rates fail to establish in new ecosystems. Strong dispersal ability does not guarantee survival after arrival. Because of this, traits like rapid growth are considered risk factors, not definitive predictors.

2. Context matters more than traits alone

Whether a species becomes invasive depends heavily on environmental and ecological context, including the disturbance levels (urbanization, agriculture, deforestation), climate compatibility, availability of empty niches, and presence or absence of predators or pathogens. For instance, the invasive plant Kudzu thrives in disturbed environments in the southeastern United States but is far less aggressive in other regions.

3. Introduction pressure is a major predictor

One of the strongest predictors of invasion is actually propagule pressure—the number of individuals introduced and how often introductions occur. Species repeatedly introduced through trade or transport have many more chances to establish.

4. Some invasive species lack "classic" traits

Not all successful invaders are fast-growing or highly reproductive. Some succeed due to other mechanisms such as chemical interference (allelopathy), escape from natural enemies, and mutualistic relationships with microbes or pollinators. For example, Garlic mustard invades forests partly by disrupting soil fungi that native plants depend on.

5. Modern invasion ecology uses multi-factor models

Because traits alone are insufficient, researchers often evaluate invasiveness using multiple factors together. These include species traits, habitat disturbance, propagule pressure, climate matching, and ecological interactions.

From a cross-ecosystem perspective, water hyacinth exemplifies how invasive plants exploit resource pulses, disturbance regimes, and biotic resistance gaps irrespective of habitat type. Its success highlights the importance of functional traits rather than ecosystem classification in understanding invasion dynamics. Consequently, management approaches that focus solely on species removal without addressing underlying drivers such as nutrient loading and habitat disturbance are unlikely to achieve long-term control.

References

A Tale of Two Invaders: Tree of Heaven and Spotted Lanternfly. (n.d.). *Integrated Pest Management*. https://www.canr.msu.edu/resources/a-tale-of-two-invaders-tree-of-heaven-and-spotted-lanternfly.

Ademola, I. O., Akanbi, A. I., & Idowu, S. O. (2005). Anthelmintic activity of *Leucaena leucocephala* seed. *Pharmaceutical Biology, 43*(7), 599–604.

Ahmed, T., Hoque, A. T. M. R., & Hossain, M. K. (2008). Allelopathic effects of *Leucaena leucocephala* leaf litter on some forest and agricultural crops grown in nurseries. *Journal of Forestry Research*, *19*(4), 298–302. https://doi.org/10.1007/s11676-008-0051-7.

Alabi, D. A., & Alausa, A. A. (2006). Evaluation of the mineral and phytochemical contents of *Leucaena leucocephala*. *World Journal of Agricultural Sciences*, *2*(1), 115–118.

Alcalde, G.T., & Buot, I.E. (2023). *Mikania micrantha Kunth (Asteraceae) as a potential invasive in terrestrial and wetland ecosystems*. J. Wetlands Biodiversity 13: 69-82.

Alemán-Ramirez, J. L., Okoye, P. U., Torres-Arellano, S., Mejía-Lopez, M., & Sebastian, P. J. (2022). A review on bioenergetic applications of *Leucaena leucocephala*. *Industrial Crops and Products*, *182*, Article 114847. https://doi.org/10.1016/j.indcrop.2022.114847.

Ali Khan, M., El-Kersh, D. M., Islam, M. S., Ara Khan, S., Kamli, H., Sarkar, C., Bhuia, M. S., Islam, T., Chandra Shill, M., Gobe, G. C., Sönmez Gürer, E., Setzer, W. N., Sharifi-Rad, J., & Torequl Islam, M. (2023). *Mikania micrantha* Kunth: An Ethnopharmacological Treasure Trove of Therapeutic Potential. *Chemistry & Biodiversity*, *20*(11), e202300392. https://doi.org/10.1002/cbdv.202300392.

Ashley Kulhanek, & Kulhanek, A. (n.d.). *Plant introductions: Meet invasive species common barberry and Japanese barberry.* BYGL. https://bygl.osu.edu/node/1726 JapaneseBarberryBCP.pdf.

Auld, B. A., & Martin, P. M. (1996). The autecology of *Mimosa invisa* Mart. Ex Colla in the humid tropical lowlands of Papua New Guinea. Australian Journal of Agricultural Research, 26(4), 689–702.

Autumn olive: Your invasive, seedy neighbor. The Nature Conservancy. (n.d.). https://www.nature.org/en-us/about-us/where-we-work/united-states/indiana/stories-in-indiana/autumn-olive/

Azeemoddin, G., Jagan Mohan Rao, S., & Thirumala Rao, S. D. (1988). Studies on the composition of *Leucaena leucocephala* oil. *Journal of Food Science and Technology*, *25*, 158.

Bahadur, S., Verma, S. K., Prasad, S. K., Madane, A. J., & Maurya, S. P. (2015). Eco-friendly weed management for sustainable crop production: A review. *Journal of Crop and Weed*, *11*, 181–189.

Banerjee, A. K., Mukherjee, A., Guo, W., Liu, Y., & Huang, Y. (2019). Spatio-Temporal Patterns of Climatic Niche Dynamics of an Invasive Plant *Mikania micrantha* Kunth and Its Potential Distribution Under Projected Climate Change. *Frontiers in Ecology and Evolution*, *7*, 477754. https://doi.org/10.3389/fevo.2019.00291.

Barker, J. (2018). *Effects of white-tailed deer and invasive garlic mustard on native tree seedlings in an urban forest*. McMaster University. https://macsphere.mcmaster.ca/handle/11375/24171.

Belz, R. G. (2007). Allelopathy in crop/weed interactions—An update. *Pest Management Science*, *63*(4), 308–326. https://doi.org/10.1002/ps.1347.

Biological and integrated control of water hyacinth, *Eichhornia crassipes* (pp. 8-20). ACIAR Proceedings.

Brazier, D. A., Brooks, S. J., Warren, C., & Gough, K. L. (Year). *Controlling Leucaena leucocephala with encapsulated herbicides*. Tropical Weeds Research Centre, Department of Agriculture and Fisheries, Charters Towers, Queensland, Australia.

Brewbaker, J. L. (1987). Leucaena: A multipurpose tree genus for tropical agroforestry. In N. T. Vergara (Ed.), Agroforestry in the tropics (pp. 39–62). Southeast Asian Regional Center for Graduate Study and Research in Agriculture.

Brewbaker, J. L. (2004). *Tree breeding practices: Nitrogen-fixing tree improvement and culture*. In Encyclopedia of Forest Sciences. https://www.example.com/encyclopedia forest-sciences/treebreeding.

Buckley, Y. M., Anderson, S., Catterall, C. P., Corlett, R. T., Engel, T., Gosper, C. R., Nathan, R., Richardson, D. M., Setter, M., Spiegel, O., Vivian-Smith, G., Voigt, F. A., Weir, J. E. S.,.

Burns, R. M., Honkala, B. H., & Technical coordinators. (2022). *Silvics of North America: Volume 2*. Hardwoods. United States Department of Agriculture (USDA), Forest Service, Agriculture Handbook 654. https://research.fs.usda.gov/treesearch/1548.

Campbell, S., Vogler, W., Brazier, D., Vitelli, J., & Brooks, S. (2019). Weed Leucaena and its significance, implications and control. *Tropical Grasslands–Forrajes Tropicales*, *7*(3), 280–289. https://doi.org/10.17138/TGFT(7)280-289.

Chandran, R. & Loyd, B. (2015). *Autumn Olive (ANR-IPM-15-012) West Virginia University Extension Service* https://extension.wvu.edu/files/d/d575d145-38f0-4927-8262-afe6faa136e1/autumn-olive-formatted.pdf?.

Chen, Y., Chen, F., Liu, L., & Zhu, S. (2012). Physiological responses of *Leucaena leucocephala* seedlings to drought stress.

Cheng, F., & Cheng, Z. (2015). Research progress on the use of plant allelopathy in agriculture and the physiological and ecological mechanisms of allelopathy. *Frontiers in Plant Science*, *6*, Article 1020.

Chiou, C. R., Wang, H. H., Chen, Y. J., & Grant, W. E. (2013). Modeling potential range expansion of the invasive shrub *Leucaena leucocephala* in the Hengchun Peninsula, Taiwan. *Invasive Plant Science and Management*, *6*(3), 492–501. https://doi.org/10.1614/IPSM-D-12-00057.1.

Chou, C. H., & Kuo, Y. L. (1986). Allelopathic research of subtropical vegetation in Taiwan. *Journal of Chemical Ecology*, *12*(6), 1431–1448. https://doi.org/10.1007/BF01012333.

Co's Digital Flora of the Philippines. (n.d.). *Acanthaceae*. https://www.philippineplants.org/Families/Acanthaceae.html.

Damascena, J. G., Leite, G. L. D., Silva, F. W. S., Soares, M. A., Guanabens, R. E. M., Sampaio, R. A., & Zanuncio, J. C. (2017). Spatial distribution of phytophagous insects, natural enemies and pollinators on *Leucaena leucocephala* (Fabaceae) tree in the Cerrado. *Florida Entomologist, 100*(3), 571–577. https://doi.org/10.1653/024.100.0303.

Dauda, W. P., Dantata, I. J., Adetunji, C. O., Abraham, P., Ifeanyi, U. J., Glen, E., Morumda, D., Abraham, S. E., Wabba, G. P., Ogwuche, I. O., & Ogundolie, F. A. (2023). *The effect of evaporative coolant structure on ornamental plants*. In Elsevier eBooks (pp. 209–235). https://doi.org/10.1016/b978-0-323-89864-5.00011-4.

Day, M. D., Broughton, S., & Hannan-Jones, M. A. (2003). Current distribution and status of *Lantana camara* and its biological control agents in Australia, with recommendations for further biocontrol introductions into other countries. *CABI Reviews, 2003*. https://doi.org/10.1079/cabireviews20033155305.

Day, M. D., Clements, D. R., Gile, C., Senaratne, W. K. A. D., Shen, S., Weston, L. A., & Zhang, F. (2016). *Biology and Impacts of Pacific Islands Invasive Species*. 13. *Mikania micrantha* Kunth (Asteraceae). Pacific Science, 70(3), 257–285. https://doi.org/10.2984/70.3.1.

Day, M. D., Wiley, C. J., Playford, J., & Zalucki, M. P. (2003). *Lantana: Current management status and future prospects*. Australian Centre for International Agricultural Research (ACIAR).

Day, M. D., Wiley, C. J., Playford, J., Zalucki, M. P., & Australian Centre for International Agricultural Research. (2003). *Lantana: Current management status and future prospects*. Australian Centre for International Agricultural Research. http://www.aciar.gov.au.

De Angelis, A., Gasco, L., Parisi, G., & Danieli, P. P. (2021). A multipurpose leguminous plant for the Mediterranean countries: *Leucaena leucocephala* as an alternative protein source: A review. *Animals*, *11*(8), 2230. https://doi.org/10.3390/ani11082230.

Fang, K., Yang, A.-L., Li, Y.-X., Zeng, Z.-Y., Wang, R.-F., Li, T., Zhao, Z.-W., & Zhang, H.-B. (2023). Native plants change the endophyte assembly and growth of an invasive plant in response to climatic factors. *Applied and Environmental Microbiology*, *89*(10). https://doi.org/10.1128/aem.01093-23.

Farmer, S., Ward, J. R., Horton, J. L., & Clark, H. D. (2016, October 10). *Southern Appalachian urban forest response to three invasive plant removal treatments*. Ashville; Biology Department, University of North Carolina Asheville.

Fernandez, C., Monnier, Y., Santonja, M., Gallet, C., Weston, L.A., Prévosto, B., Bousquet-Mélou, A. The impact of competition and allelopathy on the trade-off between plant defense and growth in two contrasting tree species. Front. Plant Sci. 2016, 7, 594.

Flory, S. L., & Lewis, J. (2009). Nonchemical methods for managing Japanese stiltgrass (*Microstegium vimineum*). *Invasive Plant Science and Management*, *2*(4), 301–308.

Franke, M. E., Zipper, C. E., & Barney, J. N. (2018). Native hardwood tree seedling establishment following invasive autumn-olive (*Elaeagnus umbellata*) removal on a reclaimed coal mine. *Invasive Plant Science and Management*, *11*(3), 155-161. https://doi.org/10.1017/inp.2018.19.

Gamal-Eldeen, A. M., Amer, H., Helmy, W. A., Ragab, H. M., & Talaat, R. M. (2007). Biological properties of *Leucaena leucocephala* constituents. *Indian Journal of Pharmaceutical Sciences*, *69*, 805–811.

Gao, R., Song, Z., Zhao, J., & Li, Y. (2025).). Predicting the distribution of *Ailanthus altissima* using deep learning-based analysis of satellite imagery. https://doi.org/10.3390/sym17030324.

Gardenia.net. (2024, November 30). *Lantana camara (Bigleaf Lantana)*. Gardenia. https://gardenia.net/plant-varieties/lantana-camara-bigleaf-lantana.

Garlic mustard (*alliaria petiolata* (Bieb.) Cavara & Grande) - eddmaps distribution - eddmaps. EDDMapS.org. (n.d.). https://www.eddmaps.org/distribution/uscounty.cfm?sub=3005.

Garlic mustard (*alliaria petiolata* (Bieb.) Cavara & Grande). Invasive.org (2008, November 10). https://www.invasive.org/browse/detail.cfm?imgnum=2146038#.

Garlic mustard. Garlic Mustard | National Invasive Species Information Center. (n.d.). https://www.invasivespeciesinfo.gov/terrestrial/plants/garlic-mustard.

Gazal, R. (2025). *Lecture slides 5–18 [PowerPoint slides]*. Brightspace.

Gholami, H., Zarei, B., Zarei, M., & Moosavi, S. (2022). Integrated Management of *Imperata cylindrica*: A Review of Sustainable Control Strategies. *Journal of Agricultural Science and Technology*, *24*(1), 1-15.

Global Invasive Species Database. (2025). *Species profile: Lantana camara*. http://www.iucngisd.org/gisd/speciesname/Lantana+camara

Gopal, B. (1987). *Water Hyacinth*. Elsevier Science Publishers.

Group (ISSG), IUCN. https://www.iucngisd.org.

Gutteridge, R. C., & Shelton, H. M. (1998). Leucaena as forage: Review. *Tropical Grassland Society of Australia*, *2*, 1.

Halliday, M. J., Panjaitan, T., Nulik, J., Padmanabha, J., McSweeney, C. S., Depamede, S., KanaHau, D. K., Fauzan, M., Yuliana, B. T., Pakereng, C., et al. (2014). Prevalence of DHP toxicity and detection of *Synergistes jonesii* in ruminants consuming *Leucaena leucocephala* in eastern Indonesia. *Tropical Grasslands–Forrajes Tropicales*, *2*(1), 71–73. https://doi.org/10.17138/TGFT(2)71-73.

Harun, I., Pushiri, H.., Aiman, A., & Zulkeflee., Z. (2021). Invasive Water Hyacinth: Ecology,.

Hegarty, M. P., Schinckel, P. G., & Court, R. D. (1964). Toxic factors in *Leucaena leucocephala*. *Australian Journal of Agricultural Research*, *15*, 153– 167.

Hendrix, J. (2024, July 23). *Double trouble: tree of heaven and spotted lanternfly threaten our environment - Olmsted Parks Conservancy*. Olmsted Parks Conservancy. https://www.olmstedparks.org/double-trouble-tree-of-heaven-and-spotted-lanternfly-threaten-our-environment/

Hollis, L. (2023, March 9). *Garlic mustard threatens the rare West Virginia white butterfly*. Invasive Species Blog. https://blog.invasive-species.org/2023/03/09/garlic-mustard-threatens-the-rare-west-virginia-white-butterfly/

Holm, L.G., Plucknett, D.L., Pancho, J.V., & Herberger, J.P. (1977). *The World's Worst Weeds: Distribution and Biology*. University Press of Hawaii. (Classic, foundational reference for weed biology)

Hong, L., Liu, M.J., Shen, H., Ye, W.H., & Cao, H.L. (2010). *Reproductive allocation of biomass in Mikania micrantha populations from two habitats, [In Chinese.] J*. Wuhan Bot. Res. 28: 598–605.

Idowu, S. O., Adeyemo, M. A., & Ogbonna, U. I. (2009). Biological evaluation of Leucaena extracts. *Journal of Biological Engineering, 3*, 14.

Imperata cylindrica. Environmental Science & Policy, 115, 120-130. in water hyacinth. Aquatic Botany, 81(1), 51–67.

iNaturalist. (n.d.). *Mile-a-Minute (Mikania micrantha)*. INaturalist. https://www.inaturalist.org/taxa/165389-Mikania-micrantha.

INaturalist. Tree of heaven (Plants of Saxony) · iNaturalist. (n.d.). *INaturalist*. https://www.inaturalist.org/guide_taxa/635939.

iNaturalist. (2016, September 26). *Lantana camara (BIS-Fogo)*. https://www.inaturalist.org/taxa/105418-Lantana-camara.

International Plant Names Index. (n.d.). *Thunbergia grandiflora*.

Invasive Species Fact Sheet Pacific Islands Area. https://www.hicattle.org/Media/HICattle/Docs/thurnbergia_grandiflora.pdf.

Invasive species centre. (n.d.). *Tree of heaven*. Invasive Species Centre. https://www.invasivespeciescentre.ca/invasive-species/meet-the-species/invasive-plants/tree-of-heaven/

Invasives in your woodland: Garlic Mustard (updated 2025). University of Maryland Extension. (n.d.). https://extension.umd.edu/resource/invasives-your-woodland-garlic-mustard-updated-2025/

Ishak, M. S., Ismail, B. S., & Yusoff, N. (2016). Allelopathic potential of *Leucaena leucocephala* (Lam.) de Wit on the germination and seedling growth of *Ageratum conyzoides* L., *Tridax procumbens* L., and *Emilia sonchifolia* (L.) DC. *Allelopathy Journal*, *37*(1), 109–122.

Isler, J., Harald Bugmann, Conedera, M., & Thrippleton, T. (2023). Long-term dynamics of tree of heaven (*Ailanthus altissima*) in central European forests. *European Journal of Forest Research*, *142*(5), 1149–1166. https://doi.org/10.1007/s10342-023-01582-9.

IUCN. (n.d.). *Invasive Alien Species*. https://iucn.org/our-work/topic/invasive-alien-species.

Jose, S., & Miller, J.H. (2004). *Imperata cylindrica* (L.) Beauv.: Ecology, Management, and Journal of Research in Agronomy.

Kamara, A. Y., Akobundu, I. O., Sanginga, N., & Jutzi, S. C. (1999). Effects of mulch from 14 multipurpose tree species (MPTs) on early growth and nodulation of cowpea (Vigna unguiculata L.). *Journal of Agronomy and Crop Science*, *182*(2), 127–133. https://doi.org/10.1046/j.1439-037x.1999.00278.x.

Kato-Noguchi, H., & Kurniadie, D. (2022). Allelopathy and allelochemicals of *Leucaena leucocephala* as an invasive plant species. *Plants*, *11*(13), 1672. https://doi.org/10.3390/plants11131672.

Kurts, C. M., & Hansen, M. H. (2016) An Assessment of Autumn Olive in Norther U.S. Forests. Research Not NRS-204. U.S. Department of Agriculture, Forest Service, Northern Research Station https://www.fs.usda.gov/nrs/pubs/rn/rn_nrs204.pdf.

Kurtz, C. (2025, September 9). *Japanese barberry*. Penn State Extension. https://extension.psu.edu/japanese-barberry Japanese Barberry.

Kurtz, C. (n.d.). *Northern Research Station*. US Forest Service Research and Development. https://research.fs.usda.gov/nrs An Assessment of Japanese Barberry in Northern U.S. Forests.

Kurtz, C. M., & Hansen, M. H. (2018). *An assessment of oriental bittersweet in northern U.S.* forests (Research Note NRS-251). U.S. Department of Agriculture, Forest Service, Northern Research Station.

Lee, H., & Lee, C. (2006). Environmental factors affecting establishment and expansion of the invasive alien species of tree of heaven (*Ailanthus altissima*) in Seoripool Park, Seoul. *Integrative Biosciences*, *10*(1), 27–40. https://doi.org/10.1080/17386357.2006.9647281.

Lewis, P. (2014, May 15). *Tree of Heaven: An Exotic Invasive Plant Fact Sheet - Ecological Landscape Alliance*. Ecological Landscape Alliance. https://www.ecolandscaping.org/05/landscape-challenges/invasive-plants/tree-of-heaven-an-exotic-invasive-plant-fact-sheet/

Li, Y., Li, J., Li, Y., Wang, X., & Cao, A. (2013). *Antimicrobial Constituents of the Leaves of Mikania micrantha H*. B. K. PLoS ONE, 8(10), e76725. https://doi.org/10.1371/journal.pone.0076725.

Lowe, S., Browne, M., Boudjelas, S., & De Poorter, M. (2001). 100 of the world's worst invasive alien species: A selection from the global invasive species database, IUCN/SSC.

MacDonald, G.E., Johnson, C.M., Minogue, P.J., & Lauer, D.K. (2018). *Cogon Grass (Imperata cylindrica) Management in Florida Forest Lands*. University of Florida IFAS Extension. Publication #FOR194.

Meena Devi, V. N., Ariharan, V. N., & Nagendra Prasad, P. (2013). Nutritive value and potential uses of *Leucaena leucocephala* as biofuel—A mini review. *Research Journal of Pharmaceutical, Biological and Chemical Sciences*, 4(1), 5515–5521.

Michigan Department of Natural Resources & Michigan Natural Features Inventory. (2012, February). Invasive Species – Best Control Practices: Autumn Olive (*Elaeagnus umbellata*). https://mnfi.anr.msu.edu/invasive-species/AutumnOliveBCP.pdf.

Michigan Natural Features Inventory. (2012). *Best Control practices for autumn olive (Elaeagnus umbellata Thunb.) Michigan State University Extension.* https://mnfi.anr.msu.edu/invasive-species/AutumnOliveBCP.pdf.

Mid-Atlantic Writers. (n.d.). *SHRUBS & SUBSHRUBS*. Japanese barberry (*Berberis thunbergii*). https://www.invasive.org/alien/pubs/midatlantic/beth.htm Japanese Barberry (*Berberis thunbergii*)

Miller, J. H. (n.d.). *Round Leaf Bittersweet*. U.S. Department of Agriculture. https://www.fs.usda.gov/database/feis/plants/vine/celorb/all.html.

Missouri Department of Conservation. Autumn Olive control. Missouri Department of Conservation. https://mdc.mo.gov/trees-plants/invasive-plants/autumn-olive-control.

Mridula, Rana, M., Sharma, G., Anshumali, Rana, S., & Sharma, S. (2025). *Lantana camara*: A review on ecology, invasion and management strategies. *International Journal of Research in Agronomy*, *8*(2), 141–147. https://doi.org/10.33545/2618060x.2025.v8.i2c.2543.

Njunge, J. T., Kaholongo, I. K., & Amutenya, M. (2017). Invasiveness and biomass production of *Leucaena leucocephala* under harsh ecological conditions of north-central Namibia. *Journal of Tropical Forest Science, 29*(3), 297–304.

Royal Botanic Gardens, Kew. (2021). *Mikania micrantha Kunth | Plants of the World Online | Kew Science*. Plants of the World Online. https://powo.science.kew.org/taxon/urn:lsid:ipni.org:names:161051-2.

Sanakaran, K.V., n.d. *Mimosa Diplotricha*: Giant Sensitive Plant, Asia-Pacific Forest Invasive Species Network. https://www.fao.org/forestry-fao/

Sandoval, J., & Rodríguez, P. (2012). *Thunbergia grandiflora (Bengal trumpet) [Dataset]*. https://doi.org/10.1079/cabicompendium.117524.

Santos, B. A., Peres, C. A., & Benchimol, M. (2023). *Forest regeneration under invasion: Impacts of Leucaena leucocephala on species richness and functional diversity*. EcoEvoRxiv. https://doi.org/10.32942/osf.io/xxxxx.

Schandry, N., Becker, C. Allelopathic plants: Models for studying plant–interkingdom interactions. Trends Plant Sci. 2020, 25, 176–185.

Seeley, N. S. (2022). *Invasive autumn olive's impacts on native red oak growth rate and germination*. (Undergraduate independent studies thesis, The College of Wooster). Department of Biology, The college of Wooster. https://wooster.edu/2022/04/26/nathaniel-seeley/

Sheam, M., Haque, Z., & Nain, Z. (2020). Towards the antimicrobial, therapeutic and invasive properties of *Mikania micrantha* Knuth: a brief overview. *Journal of Advanced Biotechnology and Experimental Therapeutics*, *3*(2), 92. https://doi.org/10.5455/jabet.2020.d112.

Sheet. https://apfisn.org.

Shen, S., Xu, G., Zhang, F., Jin, G., Liu, S., Liu, M., Chen, A., and Zhang, Y. (2013). Harmful Effects and Chemical Control Study of *M.* micrantha Micrantha H.B.K. in Yunnan, Southwest China. African Journal of Agricultural research. 8(44): 5554-61.

Shrivastava, V. S. (2012). Chemical studies on *Leucaena leucocephala*. *International Journal of ChemTech Research*, *4*(3), 1038–1043.

Sladonja, B., Sušek, M., & Guillermic, J. (2015). Review on Invasive Tree of Heaven (*Ailanthus altissima* (Mill.) Swingle) Conflicting Values: Assessment of Its Ecosystem Services and Potential Biological Threat. *Environmental Management*, *56*(4), 1009–1034. https://doi.org/10.1007/s00267-015-0546-5.

Soler, J., & Izquierdo, J. (2024). The Invasive *Ailanthus altissima*: A Biology, Ecology, and Control Review. *Plants*, *13*(7), 931. https://doi.org/10.3390/plants13070931.

Stark, C. (2001). *Control of Elaeagnus Umbellata (Autumn Olive)*. University of Minnesota, Department of Forest Resources. https://conservancy.umn.edu/server/api/core/bitstreams/85274c22-c918-4391-91e8-e8169dfebb32/content?utm.

Swearingen, J. M. (n.d.). *Oriental Bittersweet*. U.S. Department of Agriculture. https://www.fs.usda.gov/database/feis/plants/vine/celorb/all.html.

Thompson, J., Fell, R., Brewster, C., Lloyd, P., & Stipes, H. (2008). *Pollination Biology of Ailanthus altissima (Mill.) Swingle (Tree-of-Heaven) in the Mid- Atlantic United States*. https://vtechworks.lib.vt.edu/bitstream/handle/10919/32586/Thompson_Thesis_Final_5-31-08.pdf.

Tree-of-Heaven. (2021). *Ohiodnr.gov*. https://ohiodnr.gov/discover-and-learn/plants-trees/broad-leaf-trees/tree-of-heaven-ailanthus-altissima.

Verma, P. R. P., & Balkishen, R. (2007). Physicochemical studies on *Leucaena leucocephala*. *Journal of Scientific and Industrial Research, 66*, 550–557.

Watson, G.A., Wong, P.W., & Narayanan, R. (1964). Effects of cover plants on soil nutrient status and on growth of Hevea. IV. Leguminous creepers compared with grasses, *Mikania cordata* and mixed indigenous covers on four soil types, Journal of the Rubber Research Institute of Malaya 18: 123-145.

West Virginia University Extension Service. (2015). *Autumn olive (Elaeagnus umbellata): Invasive species fact sheet*. West Virginia University. https://extension.wvu.edu/files/d/d575d145-38f0-4927-8262-afe6faa136e1/autumn-olive-formatted.pdf.

Whals Jr, Carpenter SR, Zabden MJV. (2016). *Invader triggers loss of ecosystem service*. https://journal.biotrop.org/index.php/biotropia/article/view/933/594.

Whitesell, C. D. (1974). *Leucaena leucocephala, leucaena*. In C. S. Schopmeyer (Ed.), Seeds of woody plants in the United States (USDA Forest Service Handbook No. 450). Washington, DC.

Widjaja, E.A., & Tjitrosoedirdjo, S.S. (1991). The development of weeds under *Dendrocalamus asper* (Schultz. F.) Back. ex Heyne plantation in Lampung, Sumatra, Indonesia, Proceedings, 13th Asian-Pacific Weed Science Society Conference Taipei, Taiwan, Asian-Pacific Weed Science Society 1: 225-227.

Wilcut, J.W., Dernoeden, P.H., & Gilliam, C.H. (1988). *Control of Imperata cylindrica (L.) Beauv*. in Agronomic and Horticultural Crops. Weed Technology, 2(1), 1-8.

Witt, A. B. R. (2010). Biofuels and invasive species from an African perspective—A review. *GCB Bioenergy, 2*(6), 321–329. https://doi.org/10.1111/j.1757-1707.2010.01061.x.

Wong, P.W. (1964). *Evidence for the presence of growth inhibitory substances in Mikania cordata (Burm.* f.) B.L. Robinson, J. Rubber Res. Inst. Malaya 18: 231-242.

Woodmansee, J. (2023, August). Identifying tree-of-heaven [Photograph]. Purdue University Extension. https://extension.purdue.edu/news/county/whitley/2023/08/identifying-tree-of-heaven.html.

Wooten, J. (2013, May 13). Techniques to Suppress Invasive Oriental Bittersweet (*Celastrus orbiculatus*) on Presque Island State Park in Erie, Pennsylvania.

West Virginia Department of Agriculture. (2020, June 29). *Sale of Japanese barberry outlawed July 1.* https://agriculture.wv.gov/sale-of-japanese-barberry-outlawed-july-1/

Www.nparks.gov.sg. https://www.nparks.gov.sg/florafaunaweb/flora/1/4/1451.

Yang, Q.H., Ye, W.H., Deng, X., Cao, H.L., Zhang, Y., & Xu, K.Y. (2005). *Seed germination eco-physiology of Mikania micrantha H.B.K., Bot.* Bull. Acad. Sin. (Taipei) 46:.

Yusuff, A. S., Lala, M. A., Popoola, L. T., & Adesina, O. A. (2019). Optimization of oil extraction from *Leucaena leucocephala* seed as an alternative low-grade feedstock for biodiesel production. *SN Applied Sciences, 1,* 357. https://doi.org/10.1007/s42452-019-0374-9.

Zhang, L. Y., Ye, W. H., Cao, H. L., & Feng, H. L. (2004). *Mikania micrantha H.* B. K. in China - an overview. Weed Research, 44(1), 42–49. https://doi.org/10.1111/j.1365-3180.2003.00371.x.

Mary Ann R. Agudilla, PhD & Aqilah Awg Abdul Rahman, PhD

05

Lessons and Insights in Management and Community Awareness

Lessons and Insights in Management and Community Awareness

When we think of invasive alien species, the questions that come to mind are straightforward: How are they introduced? What impact do they have on native species? How do we control it? However, as our students collaborated in the Collaborative Online International Learning (COIL) program, we saw more complex questions emerge. Once a species has established its roots within a foreign ecosystem, how do local communities respond? The discussion among our students revealed that management is not just as simple as eradicating the said species from its non-native ecosystem. The negative impact of introduced species on the environment, biodiversity, and economy can be observed in multiple ways. It involves considering the economic consequences, policy strategies, and public perceptions. This chapter explores, through our students' research, how communities in West Virginia, the Philippines, and Malaysia strategically manage invasive plants.

Key lessons that students learn across international borders are that most invasive species were introduced intentionally, sometimes with the most sincere of intentions, in contrast, other species were accidentally introduced, cognizant of their social and economic benefits. As reported by the students, most of the species under their review were brought for their aesthetic value as ornamental plants, including Asian bittersweet (*Celastrus orbiculatus*), Japanese barberry (*Berberis thunbergii*), Japanese honeysuckle (*Lonicera japonica*), Multiflora rose (*Rosa multiflora*), Tree of heaven (*Ailanthus altissima*), Mile-a-minute vine (*Mikania micrantha*), Sky flower (*Thunbergia grandiflora*), Lantana (*Lantana camara*), Soap Bush (*Miconia crenata*), and Billygoat weed (*Ageratum conyzoides*). These species escape cultivation and quickly colonize disturbed habitats, roadsides, fields, and pastures, and even in closed forests, and some become successfully naturalized across countries. Accidental introductions, such as Stiltgrass (*Microstegium vimineum*) and Giant sensitive plant (*Mimosa invisa*) and Creeping wood sorrel (*Oxalis corniculata*) pose ecological threats to both agricultural and forest ecosystems.

Students reported intentional introductions of species for rapid greening projects. In the United States, for example, students noted that Kudzu (*Pueraria montana*) was initially promoted for soil erosion control, but now smothers forests, requiring costly chemical and mechanical controls. Students also cited that Autumn olive (*Elaeagnus umbellate*), a shrub from East Asia, introduced into the United States for reforestation, erosion control, and wildlife habitat and food, has spread rapidly and outcompetes native vegetation in West Virginia. The aggressive growth and high seed production of mature Autumn olive thickets disrupt ecosystem processes and reduce biodiversity. A similar case can also be observed in the Philippines with the introduction, for instance, of fast-growing tree species like Mahogany (*Swietenia macrophylla*), a high-value timber species often useful in reforestation efforts for the rehabilitation of degraded areas in the country. However, it has proliferated, outcompeting native flora through rapid growth and allelopathy. Another species, Ipil-ipil (*Leucaena leucocephala*) was intentionally introduced primarily for reforestation, erosion control, fuelwood production, and livestock fodder. Nevertheless, *L. leucocephala* forms dense monospecific stands, creating competition and allelopathic effects on adjacent vegetation. The ability of Cogon grass (*Imperata cylindrica*) to thrive in degraded landscapes provides a substantial setback to reforestation efforts in the Philippines, primarily due to its aggressive growth and high flammability, which prevent native tree species from establishing. This shows that restoring ecosystems is not limited to planting trees that favor fast growth and cover, as these actions often backfire. In their analysis, the students emphasized an important point - that true reforestation initiatives require comprehensive management interventions addressing ecological complexity, biodiversity, changing climate, and the community livelihoods to attain a sustainable ecosystem restoration.

Combating invasions is significantly more costly than their prevention. Over the years, mistakes are inevitable, no matter how well-intentioned, one needs to figure out how to correct these historical mistakes. Students report shows that the management of invasive species in the Philippines, Malaysia, and West Virginia employs integrated pest management (IPM) that combines the use of biological (natural enemies), mechanical (physical), and chemical (herbicides) controls adapted to local ecosystems and their available resources. Their management strategies and approaches generally involve early detection, eradication, and rapid containment to safeguard biodiversity and ecosystems. The control of invasive plants underscores that their management is not only a biological challenge but also a policy and social concern.

In the discussion boards, the students discussed how policies may assist in removing the pressure of invasive species on the native species within their country. This has forced a shift in policy towards proactive management, though challenges in implementation remain. Policies must be as aggressive as the invasive plants they control. In the Philippines, the students referenced the National Invasive Species Strategy and Action Plan (NISSAP) 2020-2030, which emphasizes biosecurity protocols. This was described as a multidisciplinary approach that involved both science-based research and stricter enforcement. The commitment shown by the Department of Environment and Natural Resources shows that the country's administration as a whole no longer views invasive species as a small nuisance, but rather a national threat. The same scenario can be seen in Malaysia, where the focus is often on protecting commodities. For instance, Itchgrass (Rottboellia cochinchinensis) and Cogon Grass are considered a threat not only to biodiversity, but also because they threaten the yields of rice, maize, and oil palm. The State of West Virginia, through its Department of Agriculture, has banned the sale and distribution of Japanese Barberry. Adding the species to the list of noxious weeds prohibits its transportation, which controls its further spread.

An emerging theme that can be seen throughout the COIL project was the concept of adaptive management. This concept stressed the importance of critical thinking in finding a solution to the invasive species issue. If we cannot eradicate it, find a way to utilize it instead. Many invasive plants have been integrated into local uses ranging from medicinal, agricultural, and livelihood. Local communities often use these species for human and wildlife consumption, construction, fodder, nectar sources, and traditional medicines.

The Water hyacinth (*Pontederia crassipes*) provides a great case study for this. In both the Philippines and Malaysia, this floating aquatic plant is known to clog waterways, causing flooding and depleting dissolved oxygen. Poisoning the water with herbicides is not a sustainable and safe solution. As a result, communities have come up with innovative solutions, realizing the potential to turn invasive species into resources. As noted by a student, in the Philippines, local organizations are harvesting the plant to create fiber for textiles, handicraft and bags. There is also research being conducted on converting the biomass into biogas or animal feed. Nevertheless, Water hyacinth poses significant economic drawbacks. While this argument is being put forward, another student raised a very important question: If the community becomes economically dependent on the invasive species for handicrafts, do they lose the incentive to fully eradicate it? We leave this question of socio-ecological balance for our future conservationists to answer and navigate.

Through this collaboration, we learned that the response to invasive species might be tailored between countries, but it is still ecologically focused. One cannot simply go to war against nature and invasive species. A multifaceted approach is required through policies and extensive research to manage the introduction of invasive plants by strengthening quarantine protocols and the use of native species. A socially inclusive approach is increasingly observed across countries as a paradigm shift in the control of invasive species, acknowledging the role of local communities as a first line of defense.

Aqilah Awg Abdul Rahman, PhD & Rico M. Gazal, PhD

06

What We Learned from Each Other

What We Learned from Each Other

This chapter summarizes what students learned through a COIL experience on invasive plants in the United States, the Philippines, and Malaysia. Through a series of activities, students developed an understanding of biological invasion as an ecological process shaped by human activity, site conditions, and biotic interactions. Their discussions and project outputs indicate that international collaboration enhanced conceptual understanding, strengthened field-based observation, and recognized invasive plants as global problems expressed through local conditions.

Discussion Topics 1 to 5 show a clear student learning exercise from relationship building to ecological reasoning and shared responsibility. The first discussion laid the foundation for collaborative learning. Students introduced their majors and career goals, identified with their local landscapes, and signaled enthusiasm for learning with peers in other countries. By realizing that a forestry student from West Virginia and a biodiversity conservation student from Malaysia and the Philippines share the same passion for conservation, their original sense of fear of international collaboration slowly fades. This eventually creates a welcoming environment that made later scientific exchange easier.

Building on this collaborative learning foundation, students moved into ecological explanation and comparison. In Discussion Topic 2, they introduced their selected invasive species and explained what makes it invasive, linking traits such as rapid growth, reproduction, dispersal, and competitiveness to real impacts on native species and ecosystem processes. Replies often did more than agree; students asked questions, confirmed mechanisms with local knowledge, and compared similar problems across regions, reinforcing the idea that invasion outcomes depend on both species' traits and site-specific conditions. For instance, students in the Philippines and Malaysia were surprised to find that the Water Hyacinth (*Pontederia crassipes*) is also a major concern in the subtropical United States. The discussion also revealed how context changes perception. A student highlighted how Mahagony (*Swietenia macrophylla*) which is often viewed as a valuable timber resource in forestry programs is now considered as an invader to the forests in the Philippines. This allows the students to try and understand the complexities of management ethics: When does a resource become a pest?

Discussion Topic 3 then made invasion ecology more visible and personal. Through the short spotlight videos, students practiced field identification and described habitat context, repeatedly connecting invasive presence to forest edges, disturbance, roadsides, riparian zones, farms, and freshwater settings. These videos allow their peers to observe their reality and what their ecosystem is dealing with when it comes to invasive species. Students admitted hat before this assignment, they walked past species like Japanese Stiltgrass (*Microstegium vimineum*) everyday without noticing them and its effect on the ecosystem. Peers responded by clarifying identification characteristics, comparing habitats across countries, and asking management-oriented questions, which helped students move from "what the plant is" to "why it succeeds in the region." Some students also highlighted the difference in nomenclature used for the same plant. For instance, a Malaysian student presented *Leucaena leucocephala* as Petai Belalang in Malay. Students from the Philippines immediately recognized it as Ipil–ipil, a plant that is also known in their own local ecosystem.

By Discussion Topic 4, students were moving beyond simple identification and started to think in systems. The concept map assignment required students to connect invasion routes, habitat conditions, and key biotic and abiotic drivers to explain how invasions spread and affect ecosystems. For instance, students from Malaysia diagrammed a specific relationship between *Lantana camara* and the Long-tailed Macaque (*Macaca fascicularis*), illustrating how primates consume the fruit and disperse the seeds far from the parent plant. Simultaneously, students in West Virginia also mapped a nearly identical dynamic between Japanese Honeysuckle (*Lonicera japonica*) and the White-tailed Deer (*Odocoileus virginianus*). Viewing the diagrams side by side helped students recognize native herbivores as a common mechanism facilitating invasion.

Finally, Discussion Topic 5 served as the capstone synthesis. Students shared presentations, learned from partner institutions' case studies, and reflected on how collaboration shaped their understanding of ecology, communication, and environmental responsibility. In their reflections, students noted that invasive plants are shared challenges, but their impacts and management differ by place. They also described improved communication skills and a stronger sense of stewardship beyond their own region. In their final reflection, the students described a permanent shift in perspective. They noted that they no longer view their local landscape as a static green backdrop. Now they are able to instinctively scan for threats of invasive species and identifying mechanisms such as allelopathy and wildlife dispersal roles that they previously ignored.

The pre and post COIL surveys show that students entered the module with strong awareness of invasive species, but with more limited confidence and uneven prior international experience. Before the project, most respondents reported being very or somewhat familiar with invasive species (96.61%, 57 of 59), yet only 11.86% felt very confident identifying invasive plants in their local environment, with most reporting only somewhat or slightly confident. Nearly half of respondents had not previously interacted with students from other countries in an academic setting (45.76%), which makes the high level of enthusiasm notable, as 89.83% were very or somewhat excited to participate. Students also began with moderate self-rated understanding of invasive species impacts (69.49% moderate; 30.51% high) and generally solid readiness to use online tools for collaboration.

By the conclusion of the 11-week COIL experience, the data showed a transformative shift. Students reported clear growth in both ecological understanding and the value of learning across countries. Every respondent indicated their understanding increased, with 82.46% reporting a significant increase and 17.54% reporting a moderate increase. Confidence also rose sharply, with 49.12% reporting they were very confident and 45.61% somewhat confident in identifying invasive species, suggesting that repeated exposure through discussion, spotlight videos, concept mapping, and peer examples translated passive knowledge into applied competence. Students also valued the international dimension, with about 95% rating it very or somewhat valuable, even though communication was reported most often as occasional rather than frequent, reflecting realistic time zone and scheduling constraints. Overall perceptions of the online tools were strongly positive, and the high intent to recommend the project supports the conclusion that this COIL model delivered meaningful international experience while strengthening students' ability to identify, explain, and evaluate invasive plant impacts through site comparison.

The COIL activities demonstrate that learning about invasive plants is most powerful when students connect ecological theory to local landscapes and compare those experiences across regions. By working with peers abroad, students can recognize similar invasion processes in different ecosystems, and understand how local cultural, economic, and ecological conditions shape management decisions. Overall, COIL shows strong potential as an effective way to teach invasion ecology while helping students think globally and act responsibly in their own communities.

Project Overview and COIL Timeline

Project Overview

This appendix summarizes the 11-week Collaborative Online International Learning (COIL) module titled "Understanding the Impacts of Invasive Species and Ecosystem Dynamics," implemented across Forest Ecology and related biodiversity courses at four partner institutions in three countries: Glenville State University (West Virginia, United States), the University of the Philippines Los Baños College of Forestry and Natural Resources and Southern Luzon State University (Philippines), and Universiti Tun Hussein Onn Malaysia (Malaysia).

Project Leadership and Coordination

Dr. Rico M. Gazal served as the principal module author and project lead coordinator. Faculty coordinators from the partner institutions co-planned learning outcomes and coordinated the sequence of activities to support intercultural exchange and comparative ecological inquiry.

COIL Project Coordinators

- Dr. Mary Ann R. Agudilla (SLSU)
- Dr. Amalia E. Almazol (SLSU)
- Dr. Rico M. Gazal (GSU)
- Dr. Alona C. Linatoc (UPLB)
- Dr. Aqilah Abdul Rahman (UTHM)

Participating courses

ENR 297 Biodiversity Conservation

Southern Luzon State University (SLSU), 1st Semester, AY 2025–2026.
Instructor: Dr. Mary Ann R. Agudilla

Course Description: Concepts and values of biodiversity; extent and status of biodiversity; major threats and causes of degradation and loss; on-site conservation measures; policies involved in biodiversity conservation.

FBS 161 Forest Tree Improvement

Southern Luzon State University (SLSU), 1st Semester, AY 2025–2025.
Instructor: Dr. Amalia E. Almazol

Course Description: Principles and concepts of genetics in forestry, the science of heredity, and variation in living things. It includes the history and its relevance to everyday life.

NRMT 201 Forest Ecology

Glenville State University (GSU), Fall 2025.
Instructor: Dr. Rico M. Gazal

Course Description: The study of local and regional forest communities and those environmental factors, such as light, temperature, moisture, soils, affecting their structure. This course consists of laboratory-based exercises to explain concepts of tree growth and effects of environmental factors.

FBS 136 Forest Ecology

University of the Philippines Los Baños College of Forestry and Natural Resources (UPLB), 1st Semester, AY 2025–2026.
Instructor: Dr. Alona C. Linatoc

Course Description: Forests are among the vital ecosystems that connect life and balance the environment. Protection and conservation of forest ecosystems are therefore necessary to sustain their ecological and multiple values. Understanding the dynamics and processes within forest ecosystems following a landscape approach is key to sustainable management of forests and natural resources.

This course focuses on the principles of Forest Ecology, particularly the relationships between forest plants and the abiotic factors that influence their growth and distribution. It examines how elements such as climate, soil, water, light, and temperature interact with plant communities in forest ecosystems. The course also explores the structure and dynamics of different vegetation types, including how forests are organized into layers and how plant communities change over time. A major topic is Ecological Succession, which explains the gradual development and replacement of plant communities after natural or human disturbances. In addition, the course discusses the role of Indicator Species, which are plants that reflect specific environmental conditions. Students are also introduced to ecological methods used to assess forest vegetation, such as sampling techniques and vegetation analyses that evaluate forest composition, diversity, and overall ecosystem health.

BWJ 30503 Biodiversity and Global Environmental Change

Universiti Tun Hussein Onn Malaysia (UTHM), 1st Semester, AY 2025–2026.
Instructor: Dr. Aqilah Abdul Rahman

Course Description: Since the era of industrial revolution, so much change had happened to our environment and many of these changes brought positive and negative impacts to us, humans, and to biodiversity. This course will make students understand and acknowledge the complexity of our environment, and the many ways we have influenced it. The course covers topics such as the natural processes keeping the Earth in balance and how human activities have altered these processes from waste production to pollution, from over exploitation to global warming and climate change, which then have resulted to significant loss of biodiversity and even outbreaks of illness and diseases. Finally, the course evaluates the ecological and social effectiveness of predominant strategies in the protection of environment (i.e. biotic and abiotic components, etc), locally and globally all in accordance with the national and international policies, treaties and guidelines.

Learning outcomes

By the end of the module, students:

- Describe invasive plant ecology, routes of introduction, and environmental impacts across regions.
- Compare invasive species and invasion patterns across different forest ecosystems.
- Develop a concept map connecting an invasive species to key ecosystem components such as soil, climate, biota, and physiography.
- Synthesize findings in a written report and communicate results through an oral or recorded presentation.
- Engage in intercultural academic exchange through virtual discussion, peer feedback, and reflection.

Learning activities and platform

Cross-institutional interaction occurred through a shared Brightspace course site, hosted by Glenville State University. Students engaged through discussion boards, field-based "spotlight" videos, concept map discussion, and peer feedback on recorded presentations.

Major outputs

Major outputs included a written report with a regional invasive species profile (history, native range, introduction, impacts, and control measures), original field photos, and a finalized concept map. The module concluded with presentations (live and recorded formats, as appropriate) followed by peer viewing and reflective discussion to extend learning beyond the final submission.

Scheduling note

In the original implementation, dates and times were listed in U.S. Eastern Time, with partners in the Philippines and Malaysia approximately 12–13 hours ahead depending on daylight saving time. The schedule on the next page shows the sequence of modules and core activities.

COIL Timeline

Module	Dates	Core activities and key deliverables
Module 1	Sept 15–19	Agreements and orientation; Discussion Board 1 self-introduction; Pre-project survey.
Module 2	Sept 22–26	Group formation and selection of invasive species.
Module 3	Sept 29–Oct 3	Species introduction and discussion; Discussion Board 2 basic species information.
Module 4	Oct 6–10	Work on species background for report Part 1.
Module 5	Oct 13–17	Continue report Part 1; Discussion Board 3 spotlight video on identification and habitat.
Module 6	Oct 20–24	Create invasive species concept map.
Module 7	Oct 27–31	Continue concept map development.
Module 8	Nov 3–7	Concept map discussion; Discussion Board 4 concept map.
Module 9	Nov 10–14	Finalize report including concept map; submit final report (by Nov 16 in original schedule).
Module 10	Nov 17–21	Prepare presentations; recorded video submissions for partner institutions (by Nov 30); post-project survey (by Nov 30).
Module 11	Dec 1–5	Presentations and peer viewing; Discussion Board 5 final reflection; post-project survey (by Dec 7).

Made in the USA
Las Vegas, NV
12 August 2026

2e74ec9c-2650-4e39-a3c6-fb50ae1dbe88R01